8th Ibero-American Congress on Sensors

(Ibersensor 2012)

Journal of Physics: Conference Series Volume 421

Carolina, Puerto Rico
16-19 October 2012

ISBN: 978-1-62748-298-1
ISSN: 1742-6588

Printed from e-media with permission by:

Curran Associates, Inc.
57 Morehouse Lane
Red Hook, NY 12571

Some format issues inherent in the e-media version may also appear in this print version.

Printed by Curran Associates, Inc. (2013)

For permission requests, please contact the Institute of Physics
at the address below.

Institute of Physics
Dirac House, Temple Back
Bristol BS1 6BE UK

Phone: 44 1 17 929 7481
Fax: 44 1 17 920 0979

techtracking@iop.org

Additional copies of this publication are available from:

Curran Associates, Inc.
57 Morehouse Lane
Red Hook, NY 12571 USA
Phone: 845-758-0400
Fax: 845-758-2634
Email: curran@proceedings.com
Web: www.proceedings.com

TABLE OF CONTENTS

012001 Influence of Grain Size on AZO Ceramic Synthesis 1
C Vera, S Maioco, N Rajchenberg, R Aragón

012002 Electrospun Nanobridges Towards Self-heated Gas Sensors with Enhanced Sensitivity 5
C J Camargo, A Melendez, J Robles, J Esteve, I Ramos

012003 Composites in Small and Simple Devices to Increase Mixing on Detector Surfaces 11
L F Hernandez, R R Lima, A R Leite, E R Fachini, M L P Silva

012004 Electrospun Conducting Polymer Nanofibers As the Active Material in Sensors and Diodes 20
N J Pinto

012005 Effect of Ethanol Concentrations on Few Layer Schottky Graphene Transistors 26
L E G Armas, M F G Huila, M Pojar, H E M Peres, F J R Fernandes, M A Valle, K Araki, H E Toma, A D Santos, A C Seabra

012006 Sensors Based in Suspended Gate Field Effect Transistors for pH Measurements 31
B da S Rodrigues, A L Siarkowski, O De Sagazan, S Crand, T Mohammed-Brahim, N I Morimoto

012007 Electrochemical Characterization of Poly-(3,4 Propylene-dioxythiophene) Pseudo-capacitor 35
H Sahoo, J W Park, R A Cardona, J Santiago-Avilés

012008 Synthesis, Characterization and Use of Ru-Fc Intercalation Complex As an Electrochemical Label for the Detection of Pathogen-DNA 42
M Díaz-Serrano, A Rosado, D Santana, E Z Vega, A R Guadalupe

012009 FIA-automated System Used to Electrochemically Measure Nitrite and Its Interfering Chemicals Through a 1-2 DAB / Au Electrode: Gain of Sensitivity at Upper Potentials 47
F L Almeida, S G dos S Filho, M B A Fontes

012010 Fabrication of Phosphorus Doped Polysilicon Thin-film Strain Gauges Using a 50 Microns Silicon Substrate Thickness 55
A L Siarkowski, B S Rodrigues, N I Morimoto

012011 Photodetector with Integrated Optical Thin Film Filters 61
M A Vieira, M Vieira, P Louro, V Silva, A S Garção

012012 3-D LTCC Microfluidic Device As a Tool for Studying Nanoprecipitation 68
J N Schianti, N P N Cerize, A M Oliveira, S Derenzo, M R Góngora-Rubio

012013 Nanofibers and Thin Films As a Selective Membrane for Sensors and MicroTAS 73
A N R da Silva, M L P da Silva, E R Fachini

012014 HSENSE: A High Performance Framework for Distributed Weather Sensor Networks 80
J A Ortiz, N G Santiago, J G Colom-Ustáriz

012015 Portable Measurement System for FET Type Microsensors Based on PSoC Microcontroller 85
D G Fernández, A Blanco, A Durán, C Jiménez-Jorquera, O A-de Fuentes

012016 Determination of Heavy Metals Contamination Using a Silicon Sensor with Extended Responsive to the UV 90
M Aceves-Mijares, J M Ramírez, J Pedraza, S Román-López, C Chávez

012017 Seasonal Effects in Arsenic Levels in Drinking Water in the Lagunera Region 96
A U Aguilar-Muñiz, F Valdés-Perezgasga, G G García-Vargas

012018 Measurement System for an In-vitro Characterization of the Biomechanics and Hemodynamics of Arterial Bifurcations 101
D Suárez-Bagnasco, G Balay, L Cymberknop, R L Armentano, C A Negreira

012019 Electrodes for Bio-application: Recording and Stimulation 109
M B A Fontes

012020 Electrochemical Magneto-immunosensing of Salmonella Based on Nano and Micro-sized Magnetic Particles 116
D Brandão, S Liébana, S Campoy, P Cortés, S Alegret, M I Pividori

012021 Development of an Electrochemical Biosensor for the Detection of an ADP-ribosylating Toxin, Exo a from Pseudomonas Aeruginosa 123
Y Enríquez, Y Negrón, M Navarreto, A R Guadalupe

Author Index

PREFACE

The 8th Ibero-American Congress on Sensors (IBERSENSOR 2012) was held in Carolina, Puerto Rico on 16–19 October 2012. IBERSENSOR is a forum of the Spanish and Portuguese speaking scientific community, working in the fields of sensors of every possible kind and their applications. Previous conferences in the series were successfully carried out in La Habana, Cuba (1998); Buenos Aires, Argentina (2000); Lima, Perú (2002); Puebla, México (2004); Montevideo, Uruguay (2006); Sao Paulo, Brasil (2008) and Lisboa, Portugal (2010).

IBERSENSOR 2012 participants included researchers from eleven countries in the Americas and Europe, in particular young men and women. The conference was organized and sponsored by the Partnership for Research and Education in Materials (NSF-DMR-0934195) a collaborative program between the University of Puerto Rico at Humacao (UPRH) and the University of Pennsylvania (PENN) Materials Research Science and Engineering Center, sponsored by the USA National Science Foundation (NSF). Other sponsors included the Center for Advanced Nanoscale Materials of the University of Puerto Rico at Río Piedras and the Nano/Bio Interface Center (NBIC) at PENN.

The Proceedings of IBERSENSOR 2012 include a selection of 21 research papers in the areas of Materials and Processes for Sensor Development, Nano-Sensors, Chemical Sensors, Mechanical Sensors, Optical Sensors, Wireless Sensors, Sensor signal conditioning and Instrumentation, Microfluidic Devices, and Biomedical and Environmental Applications.

Editors
Idalia Ramos University of Puerto Rico at Humacao, Puerto Rico
Jorge J Santiago-Avilés University of Pennsylvania, USA

Logos

Ibero-American Congress on Sensors (Ibersensor)

Main Sponsors

Partnership for Research and Education in Materials (PENN-UPRH-PREM) University of Puerto Rico at Humacao

USA National Science Foundation

Other Sponsors

Center for Advanced Nanoscale Materials (CNM), University of Puerto Rico, Río Piedras

Nano/Bio Interface Center, University of Pennsylvania

Peer review statement

All papers published in this volume of *Journal of Physics: Conference Series* have been peer reviewed through processes administered by the proceedings Editors. Reviews were conducted by expert referees to the professional and scientific standards expected of a proceedings journal published by IOP Publishing.

Influence of grain size on AZO ceramic synthesis

C Vera[1,3], **S Maioco**[1], **N Rajchenberg**[1] **and R Aragón**[1,2]
[1]Laboratorio de Películas Delgadas. Facultad de Ingeniería. Universidad de Buenos Aires. Paseo Colón 850. (1063) Ciudad Autónoma de Buenos Aires. Argentina
[2]CINSO-CITEDEF-CONICET-UNSAM. Juan Bautista Lasalle 4397. Villa Martelli. Buenos Aires. Argentina

Email: cvera@fi.uba.ar

Abstract. Aluminum doped zinc oxide (AZO) is a transparent conducting oxide (TCO) with oxygen vacancy mediated carrier density and mobility. Due to low oxygen self-diffusivity, the influence of thermal reducing treatments is strongly dependent on grain size. Electrical resistivity and Seebeck coefficient measurements suggest that mobility changes prevail to increase conductivity by reducing atmosphere anneal.

1. Introduction

Transparent conducting oxides (TCO) are semiconductors ($E_g > 3$ eV) with electrical resistivities in the order of milliohm.cm and optical transmittances above 80% in the visible spectrum. Their principal applications involve electrodes for liquid crystal displays (LCD), as well as various other optoelectronic devices. The use of indium oxide doped with Sn (ITO) or Zn (IZO) [1] is widespread for this purpose but cost and availability considerations have promoted the search for possible substitutes. One alternative, which has received increased attention, recently, due to low component toxicity, relative abundance and consequent low cost, as well as favorable properties is aluminum doped zinc oxide (AZO) [2]. Optoelectronic devices require TCO thin films, frequently deposited by sputtering, either from ceramic or metallic targets, in oxygen bearing reactive atmospheres. Thin film quality, in terms of electrical and optical properties depends substantially on ceramic resistivity [3] and process conditions. This work compares the electrical properties of ceramics obtained by solid-state reaction with those produced by chemical co-precipitation.

2. Experimental procedures

AZO was prepared by solid-state reaction from a mixture of oxides (Merck Analytical Grade) in stoichiometric proportions for 3.2 atomic % Al in Zn. The oxides were suspended in acetone, grounded in an agate mortar, and sieved with N°270 (52 μm) ASTM gage. The powder was pelletized in 12 mm OD discs with a hydraulic press at 98 MPa and sintered for 48 h at 1400°C in a furnace with molybdenum disilicide heating elements.

Co-precipitated samples were prepared by dissolving mixtures of the commercial alloy ZAMAC-3 and 99.99% pure metallic Zn in nitric acid (65% m/m). They were subsequently treated with NH_3 (28% m/m) maintaining a pH between 6 and 7, in keeping with optimal minimum solubility [4].

The sample morphology was examined by scanning electron microscopy and the wurtzite structure

[3] To whom any correspondence should be addressed.

corroborated by X-ray diffraction.

Controlled atmosphere reducing anneals were undertaken with reactive oxygen fugacity buffering mixtures of CO/CO_2 supplied by MKS 1159 mass flow controllers in a Lindberg tubular reactor furnace fitted with a quenching rod assembly, which provided for sample cooling in the cold end of the reactor preserving the integrity of the annealing atmosphere. The codes S6-0h, S6-2h and S6-4h identify representative samples obtained by solid-state reaction annealed at 1000°C and $p_{O_2} = 10^{-16}$ atm for the indicated times, whereas Z11-0h, Z11-2h and Z11-4h are corresponding identifiers for co-precipitated samples.

Electrical measurements were made on prisms 1.2 cm x 0.25 cm x 0.2 cm carved from the sample pellets fitted with 100 nm thick sputtered aluminium electrodes, which supplied the contact for 40 gauge copper-constantan thermocouples. The D.C. electrical resistivity was measured by the Kelvin method and the Seebeck coefficient obtained by analogue subtraction during thermal relaxation [5] under gradients not exceeding 0.25°.

3. Results and discussion

Without further treatment, solid-state reaction samples exhibit lower resistivity than co-precipitated specimens (Table 1). Negative Seebeck 66-susanaliebana.doc
66-susanaliebana.pdfcoefficients are consistent with n-type semiconduction with a magnitude inversely proportional to carrier density but co-precipitated samples are too resistive for reliable Seebeck measurements.

Table 1. Resistivity and Seebeck coefficient for solid state reaction (S6) and co precipitated

Sample	Al/Zn At. %	Resistivity [Ω-cm]	S [μ V/K]
S6	3.2	0.43	-73.04
Z11	3.2	78700	

Since electronic conduction is sensitive to oxygen vacancy concentration [6], controlled atmosphere reducing annealings at 1000°C and $p_{O_2} = 10^{-16}$ atm, just above the zinc oxide reduction boundary were undertaken for 2 and 4 h periods. The co-precipitated specimen demonstrated extreme sensitivity to these treatments, which reduced their resistivity in five orders of magnitude within the first two hours (figure 1).

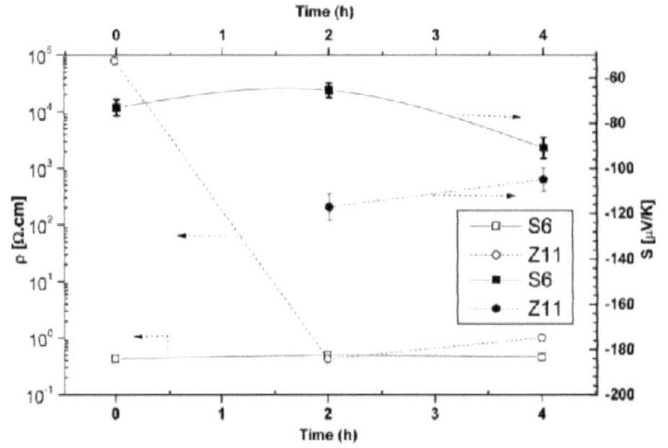

Figure 1. DC electrical resistivity and Seebeck coefficient of AZO samples prepared by solid state reaction (S6) and co-precipitation (Z11) as a function of annealing time at 1000°C and $p_{O_2} = 10^{-16}$ atm.

In stark contrast, solid-state reaction samples are essentially insensitive to the reducing annealing treatment. Scanning electron microscopy demonstrated that whereas solid state reaction produced an average grain size of 10 μm (figure 2a), co-precipitates are sub micrometric (figure 2b) even after annealing at 1400°C for comparable times.

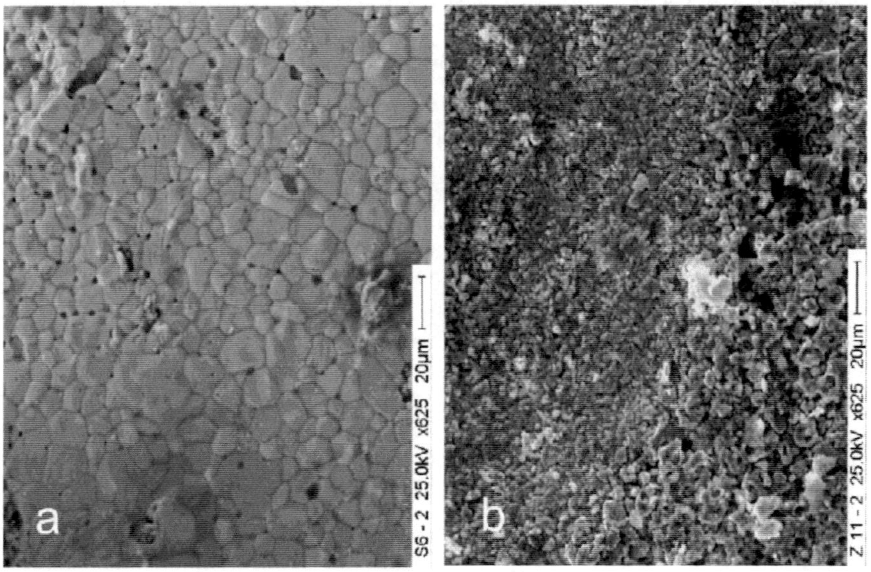

Figure 2. Scanning electron microscopy of (a) solid state reaction samples and (b) chemically co-precipitated specimens.

Preliminary solid-state reaction experiments, undertaken at 1200°C for 24 h, yielded resistivities with an order of magnitude of kΩ.cm. Higher annealing temperatures and longer sintering times contributed, not only to better homogeneity of the aluminum dopant and lower resistivity, but also to increased grain size. If the partial pressure of oxygen during the sintering process is fixed by dilution as in for instance air (log p_{O2} = -0.68), increased temperature has a reducing effect because it approaches the oxide reduction boundary leading to increased oxygen vacancy concentration and consequently lower resistivity. Cooling has the opposite effect, which requires careful attention to quenching procedures. The co-precipitated texture, of reduced grain size and higher porosity, promotes higher specific surface and consequently sensitivity to the reducing treatment. This is consistent with the low intracrystalline diffusivity of oxygen in the wurtzite structure [7].

However, the influence of reducing anneals on the Seebeck coefficient, and consequently on carrier density, is limited to changes commensurate with the experimental uncertainty of the measurements. Since conductivity is proportional to the product of carrier density and mobility, the dominant cause for the drastic decrease in electrical resistivity observed within the first two hours of reducing anneal must be attributed to increased mobility. It is also significant that, whereas resistivity is sensitively dependent on high specific surface, Seebeck coefficients reflect bulk carrier density consistently with increased contribution of grain boundary conduction to the overall resistivity [8] with smaller grain size. This behavior is well established for various semiconducting oxides, widely employed as conductimetric gas sensors [9], which respond readily to changes in surface state occupation induced by a chemical stimulus, albeit at the expense of selectivity.

The effect of reducing thermal treatments or deposition conditions for zinc oxide thin films has been noted before, although direct comparison has been hampered by the absence of specific experimental p_{O2}-T coordinates. The reactive p_{O2} buffering systems employed in this work remove this ambiguity, as well as increase the reducing character of the atmosphere with decreasing temperature, in the critical quenching stage.

4. Conclusions

Sintering is insufficient to diminish resistivity significantly in chemically co-precipitated samples. Reducing annealing treatments to optimize the metal-oxygen stoichiometry can be employed to improve the electrical properties of sub micrometric material. In practical applications of TCO that require thin films of nanometric grain size, reducing controlled atmosphere may improve processing conditions.

5. References

[1] Fortunato E, Ginley D, Hosono H and Paine D C 2007 Transparent conducting oxides for photovoltaics *MRS Bull.* **32** 242-247
[2] Minami T 2005 Transparent conducting oxide semiconductors for transparent electrodes *Semicond. Sci. Technol.* **20** S35-S44
[3] Minami T, Oda J, Nomoto J and Miyata T 2010 Effect of target properties on transparent conducting impurity-doped ZnO thin films deposited by DC magnetron sputtering *Thin Solid Films* **519** 385-390
[4] Gayer K H, Thompson L C and Zajicek O T 1958 The solubility of aluminum hydroxide in acidic and basic media at 25 °C *Can. J. Chem.* **36 (9)** 1268-71
[5] Eklundt P C and Mabatah A K 1977 Thermoelectric power measurements using analog subtraction *Rev. Sci. Instrum.* **48 (7)** 775-777
[6] Exarhos G J and Shou X-D 2007 Discovery-based design of transparent conducting oxide films *Thin Solid Films* **515** 7025-52
[7] Soares Sabioni A C, Ferreira Ramos M J and Babosa Ferraz W 2003 Oxygen diffusion in pure and doped ZnO *Materials Research* **6 (2)** 173-178
[8] Swartz C H 2012 Transport and surface conductivity in ZnO *J. Mater. Res.* **27** 2205-2213
[9] Comini E 2005 Metal oxide nano-crystals for gas sensing *Anal. Chim. Acta* **568** 28-44

Acknowledgment
This work was partially supported by UBACyT grant 20020100100968

Electrospun nanobridges towards self-heated gas sensors with enhanced sensitivity

C J Camargo[1,2,3], A Melendez[2], J Robles[2], J Esteve[1] and I Ramos[2]
[1] Instituto de Microelectrónica de Barcelona IMB-CNM (CSIC), 08193, Barcelona, Spain
[2] Department of Physics and Electronics, University of Puerto Rico at Humacao, Humacao, PR 00791, United States

E-mail: carlos.camargo@csic.es

Abstract. This paper reports the fabrication process of ZnO and GaN nanobridges by electrospinning, to be used for gas sensing applications. The main purpose is to produce single suspended fibers with their surfaces completely exposed to the gas for an enhanced sensitivity when compared with sensors fabricated with structures deposited on a substrate. In addition, the isolation from the substrate is highly favourable for a well-controlled current setting, taking advantage of self-heating effects to set appropriate temperatures for gas sensing. The current status of development of the nanobridge-based sensors is described in addition to a primary electrical characterization and the ongoing work involving the in-parallel fabrication.

1. Introduction

Semiconductor-based gas sensors have wide applications in medical diagnosis, environmental sensing and control applications. Recent research has been focused on nanostructured devices and new sensing concepts coming from nanoscience and technology. The reduction in grain size of nanostructured sensing materials, enables very high surface-to-volume ratios and therefore a wider area susceptible to gas exposures.

Gas sensing is possible thanks to the convection of electrical conductivity due to surface reaction such as oxidation or reduction caused by different gases. In detail, when the sensing material is exposed to a gas, a depletion of carriers is experienced in its grain boundaries and its properties are temporarily modified. Thus, the sensor response is highly dependent of the surface-to-volume ratio of materials because of the active centers and the defects existing on the surface layer of the materials. [1]

In comparison to conventional gas sensing devices such as those using bulk or thin films, one-dimensional nanostructures such as electrospun nanofibers have higher gas sensitivity due to their ultra-high surface-to-volume ratio. In the form of fibers, the transport properties could be strongly modified by reactions at grain boundaries and depletion of carriers. In this regard, Huang, *et al*, made a comparison between a SnO_2 thin film and SnO_2 nanorods obtained from a plasma-treated thin film. The sensivity was higher for nanostructures in a factor of 8. [2] Since this property was demonstrated, other gas sensors based on nanostructures have been studied. Different sensing metal oxides such as

[3] To whom any correspondence should be addressed.

Published under licence by IOP Publishing Ltd

SnO_2, ZnO, WO_3, even conductive polymers like PEDOT-PSSA, have been used for sensing purposes of several gases including ammonia, hydrogen, nitrogen oxides, hydrogen sulphide, and CO, just to mention a few. [3-7]

Nevertheless, the nanostructures also carry on some disadvantages due to their dimensions, for example reduced lifetime. This problem is originated by the temperature gradients in the structure due to self-heating while it operates electrically [8, 9]. This issue is overcome by applying a well-controlled current value to an individual fiber [10], which allows setting an appropriate temperature to operate the fiber as gas sensor. This was already described by Prades *et al* using SnO_2 nanowires deposited by CVD. Their system exhibited responses nearly identical to those obtained with integrated microheaters, demonstrating the feasibility of self-heated gas sensors with ultralow power requirements. [7]

Our proposal is detecting the gases or vapours using individual electrospun nanofibers suspended as bridges. It is well known that electrospinning is an efficient, relatively simple and low cost way to produce nanofibers of several materials like metal oxides and conductive polymers, which have been already studied as single fibers for gas sensing [11, 12]. In comparison to these studies, where the fibers are deposited and sintered (in the case of metal oxides) over a substrate and usually are belt-shaped, our work leads to the enhancement of gas sensitivity since the exposed area is probably doubled. In addition, a suspended fiber avoids the significant thermal losses to the support, which allows a higher effectiveness of the power released in the structure. [9]

In order to prove the concept, we report the fabrication process of Zinc Oxide (ZnO) and Gallium Nitride (GaN) nanobridges. We also discuss some aspects related to their characterization, structural and electrical characterization. Finally, some guidelines for ongoing and future work are proposed.

2. Experimental procedures

Two stages are necessary to fabricate nanobridges by electrospinning. Firstly, the micromachining and preparation of substrates where the fibers are collected. Secondly, the deposition, sintering and nitration (only for GaN) of fibers.

Substrates were prepared from conventional 500-μm-thick silicon wafers. They are firstly covered by 400 nm of thermal SiO_2, deposited as electrical isolation layer. Then, the oxide and the substrate were etched by reactive ion etching (RIE) in order to make 10 μm-height structures with rectangular and squared patterns (see grid pattern in Figure 1). A photoresist of 2 microns was enough to serve as mask during RIE. The gaps between high-relief patterns are in the range from 5 to 50 μm. A similar process was performed to fabricate a uniform array of squared 'mushrooms' of 3x3 μm^2 separated 3 μm between them. SEM images of the substrates with some fibers deposited on them are shown in Figure 2.

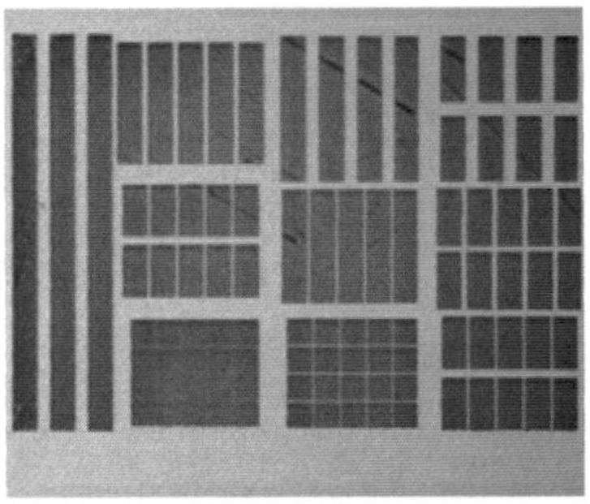

Figure 1. Grid pattern used to perform the array of high-relief structures.

Figure 2. SEM images of micromachined structures with some suspended fibers on them.

Once the substrates are prepared, GaN and ZnO nanofibers are deposited on them, using conventional electrospinning technique. First, a polymeric solution based on cellulose acetate (CA) was prepared to be used as carrier of the Zinc Nitrate ($Zn(NO_3)_2$) and Gallium Nitrate. The CA is previously dissolved in a 1:2 molecular weight ratio mixture of dymethylacetamide (DMA) and Acetone. CA is added to obtain a 20% wt solution.

In parallel to this, a 30% wt $Zn(NO_3)_2$ solution is prepared, using acetone as solvent (for ZnO), and 20% wt gallium nitrate solution was made using DMA instead acetone (for GaN). Each solution was mixed several hours with the polymeric solution in a volume ratio of 1:2. More details about precursor solution are presented in [13].

After deposition, proceeds the calcination of polymer and sintering of fibers. ZnO bridges are obtained by sintering the fibers at 600 °C in air for 3 hours, with a previous temperature ramp of 5 °C/min. Higher ramps were tried, but efficiency of bridges performing was low. Probably the forces and mechanics involved during sintering needs slow thermal variation. It is an issue that should be studied in order to improve the efficiency looking forward commercial nanofiber-based devices.

The GaN fiber is obtained by two stages: calcination/sintering and nitration. The first is made in nitrogen atmosphere at 500 °C during 3 hours with a previous ramp of 2.5 °C/min. Then, the atmosphere is changed to ammonia flowing at 40 sccm, the temperature is increased up to 900 °C at the same ratio and is hold for 3 hours. Once this phase has finished, ammonia atmosphere remains during cooling, which is controlled to be no higher than -5 °C/min.

3. Results and discussion

Electrospun nanobridges were successfully deposited as shown in figures 3 and 4. SEM characterization allows observing that ZnO bridges are highly granulated, which is favourable for gas sensing applications since the gas penetrates the structure and even larger area is exposed, however it could also make the bridges structurally weak and easy to break. The structures present a mean diameter of 700 nm, thinner fibers can be obtained if sintering time is longer or other parameters of electrospinning are controlled.

On the other hand, GaN bridges are thinner since the sintering process is longer and mainly because the temperature is higher. The mean diameter of fibers is around 150 nm, although 70-nm-fibers were observed. Although the GaN fibers present some granulation, in comparison to ZnO structures it is very low. The GaN fibers look more consistent and mechanically stronger, which make them suitable for longer bridges.

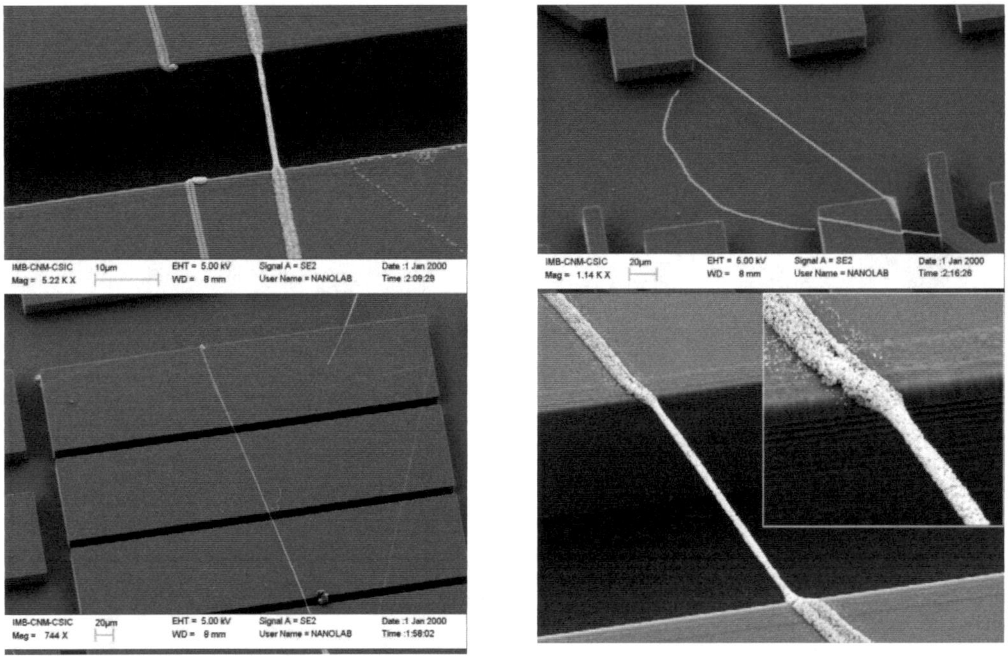

Figure 3. Electrospun ZnO nanobridges with lengths from 10 to 150 μm. A broken fiber is depicted to show that bridge performing is not a trivial process and temperature ramp affects sintering. Also paths of polymer fibers where the zinc was not sintered can be observed in the images. On right-bottom, a detailed image shows the high granulation of fibers.

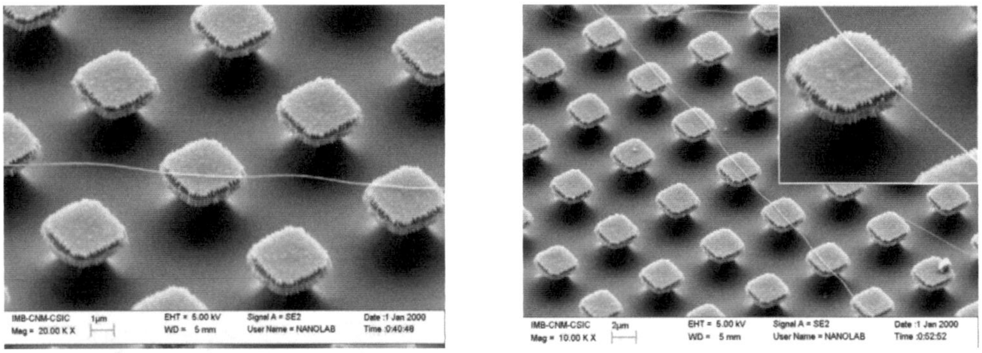

Figure 4. (Top) Electrospun GaN nanobridges deposited on mushrooms. The fibers are visibly thinner than ZnO ones. (Bottom) A detailed image shows the granulation of GaN fibers. Although it is granulated, the appearance is of a solid wire.

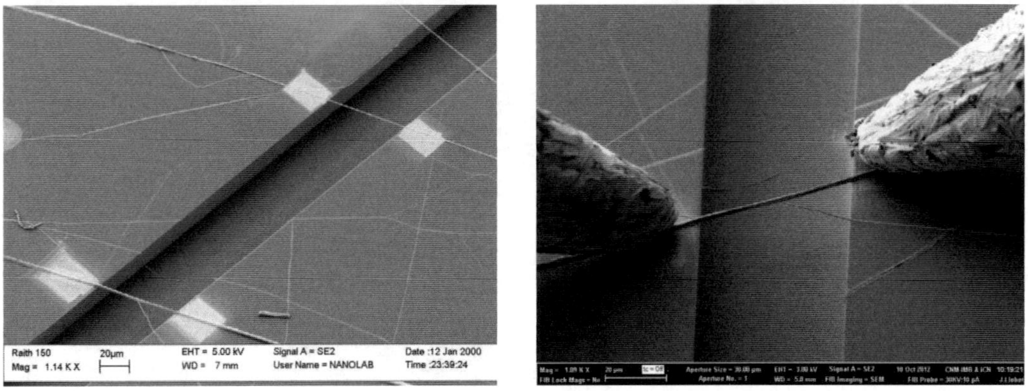

Figure 5: SEM images showing two basic setps for electrical characterization: metal pads deposition (left) and biasing (right).

The electrical characterization was carried out using a source-measurement unit (SMU) integrated to a focused ion beam (FIB) chamber. In order to apply the voltage, metal pads are necessary. By FIB-assisted deposition of a thin film of Pt, the fibers are ready to be biased as figure 5 shows. Pads have dimensions of approximately 20x30 μm^2 and were deposited the closest possible to the edge of the high-relief structure for ensuring that results come exclusively from bridge. Then, tips connected to SMU are located on these pads and measurement is performed.

Figure 6 depicts the I/V curve of a single ZnO bridge with a diameter of 1.5 μm and length of 54.7 µm. Other fibers were tested and as can be expected, the resistance varies according to the their dimensions.

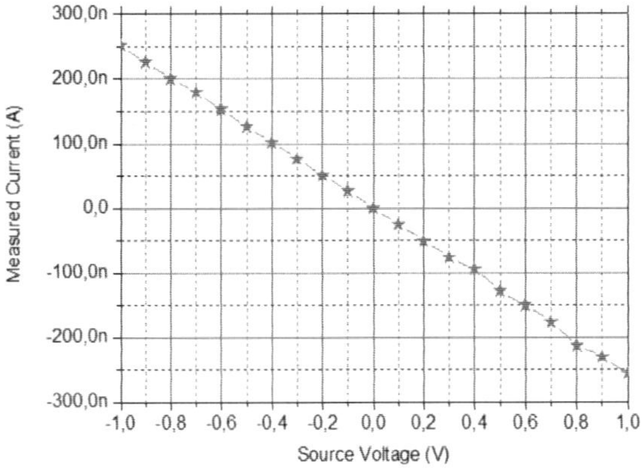

Figure 6. I/V curve of a single ZnO bridge. The structure has an ohmic (resistor) behaviour, with resistance value of 4 MΩ.

4. Ongoing work

Single and suspended electrospun nanofibers were produced. The next steps involve the characterization of fibers as gas sensors. This process is preceded by the study of their thermal behaviour to test their self-heating capabilities and guarantee the temperature control by current-biasing. However, biasing each single bridge using FIB-assisted metal deposition is a highly time/money demanding technique. For this reason, the in-parallel metallization is planned to be done by stencilling. Details about this proposal can be found in [14].

5. References

[1] Ponzoni A, Comini E and Sberveglieri G 2006 *Appl. Phys. Lett.* **88** 203101
[2] Huang H, Tang Q K, Lee Y C, Tran T D and Tse M S 2005 *Appl. Phys. Lett.* **87** 163123
[3] Wang B, Zhu L F, Yang Y H, Xu N S and Yang G W 2008 *J. Phys. Chem. C* **112** 6643
[4] Zhu C L, Chen Y J, Wang R X, Wang L J, Cao M S and Shi X L 2009 *Sens. Actuators B* **140** 185
[5] Pinto N J, Rivera D, Melendez A, Ramos I, Lim J H and Johnson A T C 2011 *Sens. Actuators B* **156** 849
[6] Rout C S, Hegde M and Rao C N R 2008 *Sens. Actuators B* **128** 488
[7] Prades J D, Jimenez-Diaz R, Hernandez-Ramirez F, Barth S, Cirera A, Romano-Rodriguez A, Mathur S and Morante J R 2008 *Appl. Phys. Lett.* **93** 123110
[8] Pop E, Mann D A, Goodson K E and Dai H 2007 *J. Appl. Phys.* **101** 093710
[9] Strelcov E, Dmitriev S, Button B, Cothren J, Sysoev V and Kolmakov A 2008 *Nanotechnology* **19** 355502
[10] Hernandez-Ramirez F, Prades J, Tarancon A, Barth S, Jimenez-Diaz R, Pellicer E, Rodriguez J, Romano-Rodriguez A, Morante J Spannhake J and Mueller G 2007 *Nanotechnology* **18** 495501
[11] Wang Y, Ramos I and Santiago-Aviles, J 2007 *IEEE Sens. J.* **7** 1347
[12] Pinto N J, Ramos I, Rojas R, Wang P and Johnson A Jr 2008 *Sens. Actuat. B-Chem* **129** 621
[13] Melendez A, Morales K, Ramos I, Campo E M and Santiago-Avilés J 2009 *9th IEEE Conf. on Nanotechnology (Genoa)* pp. 269-272
[14] Camargo C J, Esteve J, Ramos I, Campo E M and Campanella H 2011 *8th Spanish Conference on Electron Devices (Palma de Mallorca)* pp.1-4

Acknowledgments
This research is partially funded by PENN-UPRH PREM (NSF-DMR-0934195).

8th Ibero-American Congress on Sensors (Ibersensor 2012) IOP Publishing
Journal of Physics: Conference Series **421** (2013) 012003 doi:10.1088/1742-6596/421/1/012003

Composites in small and simple devices to increase mixing on detector surfaces

L F Hernandez[1], R R Lima[1,2,4], A R Leite[1], E R Fachini[3] and M L P Silva[1,4]
[1]Polytechnic School of University of São Paulo, Ave. Professor Luciano Gualberto, n. 158, Travessa 3, São Paulo, Brazil
[2]Institute of Physics, University of São Paulo, Brazil
[3]College of General Studies, University of Puerto Rico at Río Piedras, Puerto Rico
[4]Faculty of Technology of São Paulo, CEETEPS, Brazil

Email: rrlima@if.usp.br

Abstract. This work aims at three different applications for the betterment of plasma generated-composite thin films: pre-mixing, spray formation in miniaturized structures and an increase in the performance of detector surfaces. Miniaturized structures were projected, simulated with FEMLAB® 3.2 software and then constructed. Clustered films made from tetraethoxysilane (TEOS) and nonafluoro(iso)butyl ether (HFE®) precursors were deposited on silicon, acrylic and quartz substrates for different kinds of film characterization/or in the projected structures. Physical and chemical characterization guided the selection of best films previous to/after UVC exposure. The active surfaces (plasma-deposited films) in structures were modified by UVC exposure and then tested. The applications include pre-mixing of liquids and/or spray formation, best results being obtained with surface covered by derivative-HFE films, which acted as passivation layers. Preliminary results show good humidity sensing for TEOS-derivative films.

1. Introduction

TEOS (tetraethoxysilane) is the most common reagent used in the production of silicon oxide layers. The usual dry process includes plasma deposition, normally performed at high oxygen levels and high temperature. However, recently TEOS plasma polymerization at room temperature has come to the users attention, principally for flexible electronic structures and biosensors. In this case, low temperature is usually required to avoid degradation of the polymeric substrate [1]. In the absence of oxygen, TEOS polymerization leads to a hydrophobic silicone-like structure that can adsorb volatile organic compounds (VOCs) and water vapor. This polymeric chain can react when exposed to ultraviolet radiation, reactions presenting two main phases: initially carbon radicals are removed from the Si-O chain and SiOH is formed, generating a more hydrophilic surface. The second phase favors the appearance of Si-O-Si due to SiOH cross-link, restoring the hydrophobic character of the surface. If water is present, the film can suffer hydrolysis reaction forming a porous film with SiOH on the surface. Since acids or bases can catalyze this reaction, TEOS films present low resistance toward harsh environments [2]. As TEOS films are sensible to ultraviolet radiation (UV), alternating hydrophilic/hydrophobic surfaces can be created if some areas of the surface are exposed to UV. This

[4] To whom any correspondence should be addressed.

prospect can be explored in the design of microstructures for micro-unit operations [3], as pre-columns in chromatographic micro-systems, micro-reactor systems or spray formation [4]. Three-dimensional micro-channels with different walls, i.e. half-sections of hydrophobic and hydrophilic surfaces, can separate immiscible phases [5]. Even planar structures, e.g. a sealed cavity, can allow mixing of water and 2-propanol [6] or water and glycerin, a viscous fluid [7]. Nonetheless, the mixing of water-oil samples is an important challenge for a micro-unit operation, because the fluids have different viscosities and low solubility's. The low resistance of films to acidic and base solutions may hinder many of the possible applications, such as biosensors production.

Our previous work considered the tailoring of surface properties by co-deposition of HFE® (nonafluoro(iso)butyl ether) derivative plasma films: permeability to organic compounds, resistance to harsh environments or ultraviolet exposure, and oleophobic behavior was achieved by changing the film deposition parameters [8,9]. TEOS/HFE co-deposition leads to a good compromise between adsorption and resistance; preliminary results pointed out that miniaturized 3D-structures, with films on the inner surface modified by UV, might improve mixing of aqueous samples [10]. Our present work shows the use of composite thin films made from TEOS and HFE precursors in: 1) miniaturized structures for gas/liquid mixing or heterogeneous drop formation (spray); 2) passive layers on detector surfaces.

2. Experimental

Plasma composite thin films were produced using tetraethoxysilane (TEOS, Merck, USA, industrial grade) and/or methyl nonafluoro(iso)butyl ether (HFE®, 3M, USA, industrial grade). The DC powered plasma equipment was home-made and is described elsewhere. Three different substrates were used: 1) silicon wafers (type P, 100, 10-20 Ω·cm, Silicon Sense, Inc., USA), for composite characterization; 2) piezoelectric quartz crystal for water interaction tests (4.096 MHz, 8.0 mm diameter and 0.8 mm thickness, Hosonic Industrial do Brasil LTDA, Brazil); and 3) acrylics, for microstructures production and tests. All other reactants were P.A. grade and deionized water was used. The formed film was characterized after every step; every deposition step was followed by a concomitant deposition on silicon substrate. The following characterization techniques were carried out: a) ellipsometry to define refractive index, b) profilometry to determine the film thickness, c) optical and/or scanning electron microscopy (SEM) to observe cluster formation and/or roughness and d) contact angle measurements (using a goniometer and 4.0 µL drops of water) to evaluate capillarity (hydrophilic) characteristics. The corresponding SEM photos were analyzed with ImageJ® software in order to determine the distribution and the clusters size as deposited or after exposing to ultraviolet light (UVC, 1.8 W/m^2, 10 cm apart). Fourier Transformed Infrared Spectroscopy (FTIR), x-ray photoelectron spectroscopy (XPS) and Raman microscopy gave information about chemical characteristics of the films. Raman microscopy information was used to get insight about the chemical structure of the clusters. Resistance toward base/acidic aqueous solutions was evaluated by optical microscopy. QCM (quartz crystal measurements) experiments were performed with the piezoelectric crystals inside a controlled climate chamber.

The fluid behavior through microstructures, whose surfaces were pristine or exposed to UVC, was simulated using FEMLAB® 3.2b and two-dimensional or three-dimensional analysis, with an incompressible Navier-Stokes model. Differences in surface properties, such as hydrophobicity, were simulated according to Santos [11]. Santos considered two main trends: walls with and without interactions with the fluid. In the first case, the surface energy is different from zero; but no interaction means that the velocity is zero on the walls. The structures were projected and designed using Solid Edge ST4, Academic version, Siemens. They are: a) 3D-micro-channels: figure 1A presents the two parts that were machined from acrylic pieces using a conventional lathe in order to produce a single micro-channel. The surfaces of these parts can or cannot be modified by a plasma film/composite that was exposed to UVC. The micro-channel was 100 µm wide, 40 µm deep and 73 cm long, i.e. it has an area of 1.4 cm^2 and a volume of 1.1x10^{-3} cm^3, being the area/volume ratio equal to 1,300. These structures were used for mixing or for spray formation purposes. For spray tests, the 3D-micro-channel was coupled with a second structure, shown in figure 1B [4]. b) The micro-reactor: in this case, the inner part of structure (figure 1A) does not have the machined micro-channel (figure 1C) which increases the role of differences in the hydrophilicity of the surface. c) Sealed cavity: a planar device was manufactured and modified by plasma films and UVC exposure (figure 1D), as proposed previously with a different plasma film [7]. This device configuration maximizes the influence on hydrophilicity or oleophobicity.

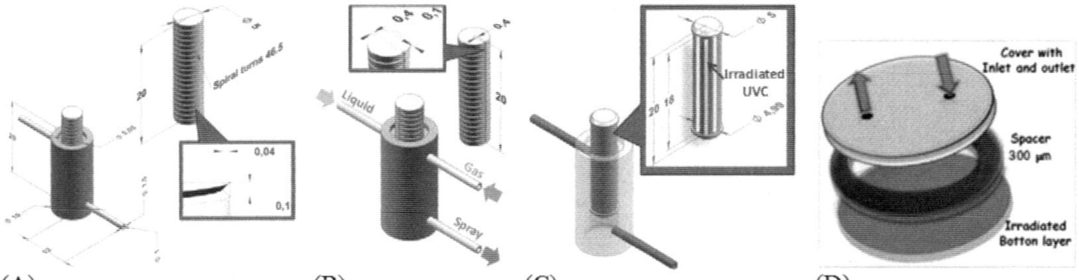

(A) (B) (C) (D)

Figure 1. Schematics of mechanical masks used to UVC exposure of (A) external and (B) internal parts of 3D structures and (C) planar structures.

The setups for insertion of nitrogen or air as gases [12] or water as liquid fluid [4] in the microstructures were described previously. The flow rate for gases was 5.0 sccm and ranged from 1.0 ml/min to 10.0 ml/min for liquids. Glycerin (650 cSt), mineral oil (viscosity of 43 cSt, Casa Americana, Brazil, tanned with Sudan black dye) or methylene blue aqueous solution (1% w/w) were used as tracers to observe fluid behaviour. Mechanical masks used for UVC exposures are represented in figure 2.

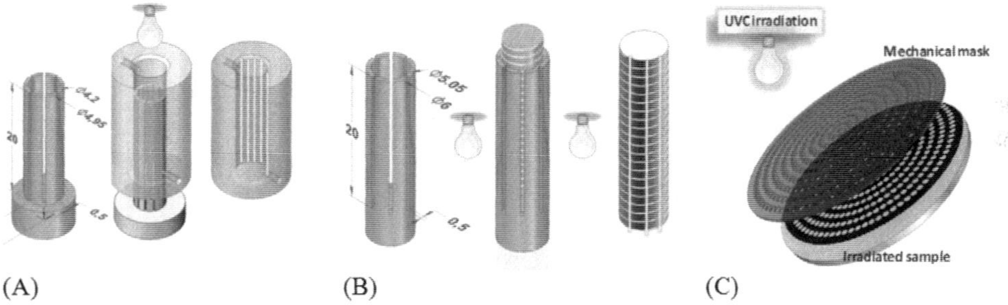

(A) (B) (C)

Figure 2. Schematics of mechanical maks used to UVC exposure of (A) external and (B) internal parts of 3D structures and (C) planar structures.

3. Results and discussion

This item presents the main characteristics of the plasma films and uses for the layers.

3.1. *The film formation*

The process of formation of plasma-composite films from TEOS or TEOS/HFE precursors and their main characteristics were described elsewhere [6,13]. The criteria for use considering the film characteristics are presented. The homemade plasma instrument made possible the production of clustered material. For films formed from TEOS and HFE, the main characteristics are:

I. Due to TEOS polymerization: a) graphite clusters can be formed and assure an adsorptive layer for VOCs; b) carbon radicals, such as OEt (etoxy), are present in the film, which suggests the presence of silicone radicals. These radicals can react under UVC conditions, leading to SiOSi and/or SiOH formation, which leads to a more chemically resistant and less hydrophobic surface, respectively; c) water nucleophilic reactions can hydrolyze OEt radicals, leading to SiOH formation.

II. Due to HFE polymerization: a) some films present numerous organic clusters, formed mainly by fluorine organic compounds, which are hydrophobic, oleophobic and resistant to UVC exposure and acid and base solutions; b) films with few organic clusters presented low water contact angles but are still more resistant to acid/base solutions than TEOS samples; c) graphite clusters can be obtained in HFE films, but only for plasma at high power conditions, i.e., high ion bombardment on the surface.

Whereas TEOS\HFE composite samples are more useful as passive layers for mixing in miniaturized structures, TEOS films are suitable for sensor structures.

Table 1 summarizes the main characteristics of pristine films produced according to the selection criteria and the results after UVC exposure. Probably the oxidation during UVC exposure decreased cluster density and cluster size. Silicone clusters can be formed during exposure (ex.: samples A, B) due to a crosslink in TEOS-derivative films. Also SiOH can be formed, which contributes to lower water contact angles (ex.: samples A, C). The changing of the water contact angle is random; high values indicating hydrophobic surfaces. The surfaces were organophilic, and they were wetted by n-hexane and 2-propanol. The refractive index did not change significantly, ranging from 1.4 to 1.6, which is characteristic of samples with high carbon content.

Table 1. Main characteristics of TEOS and TEOS/HFE films, according to the selection criteria.

For TEOS film							
	Before UVC				After UVC		
Sample type		Graphite clusters	Silicone clusters	Water contact angle (°)	Graphite clusters	Silicone clusters	Water contact angle (°)
	Refractive index	Size (μm²) / density(#/mm²)	Size (μm²) / density(#/mm²)		Size (μm²) / density(#/mm²)	Size (μm²) / density(#/mm²)	
A	1.533	0.919 / 13,300	nd	92	0.842 / 10,100	2.853 / 75	87
B	1.522	0.326 / 22,800	0.248 / 29,000	99	0.198 / 21,000	0.308 / 37,300	98
C	1.470	0.375 / 14,100	0.245 / 4,050	86	0.180 / 4,550	0.629 / 8,800	81
For TEOS/HFE film							
	Before UVC				After UVC		
Sample type		Graphite clusters	Silicone clusters	Water contact angle (°)	Graphite clusters	Silicone clusters	Water contact angle (°)
	Refractive index	Size (μm²) / density(#/mm²)	Size (μm²) / density(#/mm²)		Size (μm²) / density(#/mm²)	Size (μm²) / density(#/mm²)	
D	1.488	0.394 / 22,200	0.271 / 101,000	101	0.150 / 6,600	0.292 / 197,000	95
E	1.468	0.324 / 2,000	0.262 / 200	55	0.100 / 1,100	nd	54
F	1.429	nd	0.511 / 44,400	29	nd	0.125 / 54000	33

nd= not detected

Although deposition rates up to 20 nm/min can be achieved for these films, samples were produced at low range – 10 nm/min – because high rates require high plasma power and low precursor pressure. These plasma conditions promote high ion bombardment at the surface and, consequently, higher graphite concentration in the formed film. TEOS and TEOS/HFE films are composed of silicone-like structures with carboxyl and/or fluorine groups, respectively. Atomic concentration of oxygen is approximately 10% in all samples, as revealed by XPS, but it increases up to 17% after UVC exposure, which indicates that oxidation occurred on surface. The band for SiOSi in the FTIR spectrum was red-shifted by 20 cm^{-1} after UVC exposure, revealing that polymer chains suffered cross-linking. FTIR bands for fluorine species at 1745 cm^{-1} – 1640 cm^{-1} region were weak. This supports the results from XPS C1s deconvolution, that revealed a barely perceptible peak for fluorine species on the surface. CHn species were observed in the XPS spectra for pristine or UVC treated surfaces. Deconvolution of the C1s peak suggested the presence of graphite, whereas Raman microscopy revealed clusters formed by amorphous carbon (D band, 1300 cm^{-1} – 1500 cm^{-1}) and graphite (G band, 1600 cm^{-1}). The relative intensity of G and D bands was 0.5:1, respectively. These data differ from the corresponding information for samples obtained with plasma or photolytic procedures [14]. In summary, UVC exposure promotes several simultaneous reactions in the films, favoring crosslinks and/or oxidation of organic radicals and producing carbon clusters. Although the amount of fluorine is low, it was enough to protect the surface during 24 hours of UVC irradiation. All samples were acid and base tolerant, at least during 5 minutes of contact.

3.2. *Uses*

The main results obtained with the micro structures follow:

Figure 1 C shows the micro-reactor. The simulation of a water flow (5 ml/min) in such structures shows that the vorticity is homogeneous on all surfaces. The velocity of the fluid changes near the exit (Fig 3 A and B, respectively), i.e., a preferential path occurred. The creation of 4 hydrophobic zones changes this profile. Perturbations in the vorticity (Fig 3 C) and the velocity (Fig 3 D) occur in several points, and favor other paths. The corresponding experimental results (5 ml/min for water; 1 ml/min for mineral oil tanned with Sudan black) show: a) pristine films on the reactors surface favored one preferential path (Fig 4A); b) the UVC treated film, now with hydrophobic and hydrophilic (UVC treated area) zones, led to higher oil dispersion. The hydrophilic zone looked like an oil free area (Fig 4B).

(A) (B) (C) (D)

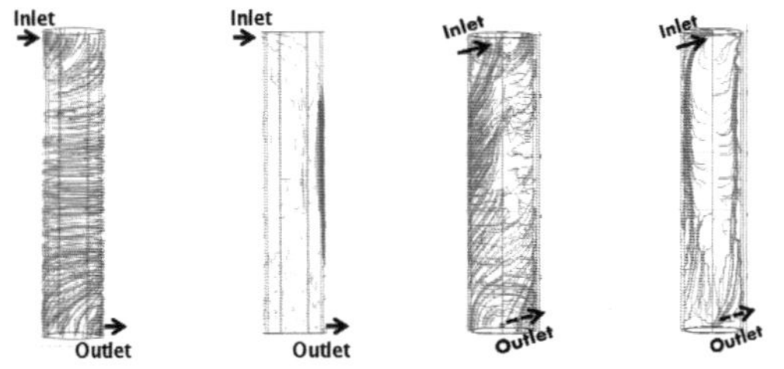

Vorticity Velocity Vorticity Velocity

Not exposed to UVC **Exposed to UVC**

Figure 3. Streamlines for vorticity (A,C) and velocity (B,D) field in a micro-reactor.

(A) (B)

Not exposed to UVC Exposed to UVC

Figure 4. Water and mineral oil flows in structures (A) without and (B) with UVC exposure.

These phenomena were enhanced in smaller dimensions; computer simulations predict the device performance. The fluid behavior on a 3D-micro-channel (Fig 1A) is similar to a chromatographic capillary column, with low vorticity and a velocity profile that does not change inside the channel [12]. On such conditions, a long and oleophobic channel full of water and oil (same conditions as figure 4) did not promote mixing. The oil fluid that travels through the channel dispersed and forms big drops with elongated shape inside the channel, as can be observed in the snapshots of figure 5A. When a micro channel with a surface modified by UVC irradiation was used, small drops (estimated volume 0.1 μL or less) were formed, and mixing is clearly observed (figure 5B). Simulations of velocity and vorticity profiles on such a micro-channel are shown in figure 5C. As the figure points out, differences in the surface hydrophilic/hydrophobic characteristics favored sudden variation in the velocity and the vorticity profiles. Similar behavior can be obtained for a gas flow (data not shown).

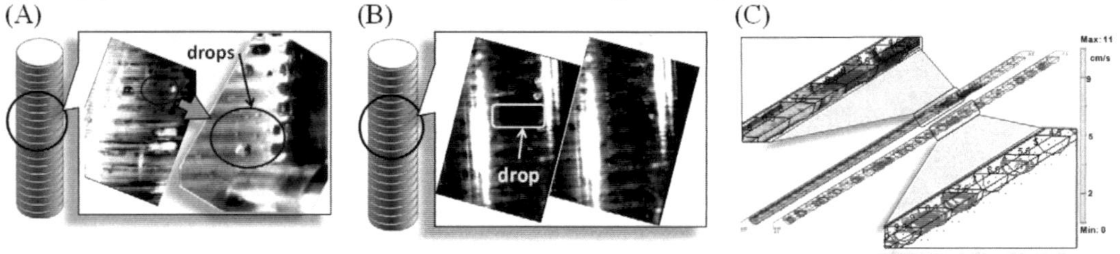

Figure 5. Experimental results: snapshots of oil interaction on a micro-channel (A) oleophobic; or (B) with hydrophilic/hydrophobic areas; (C) Simulation results: the velocity and the vorticity profiles on a micro-channel with hydrophilic/hydrophobic areas.

Two micro-unit operations [3] – mixing and spray formation – could be to advantage in using this system as a pre-mixing device. Figure 6A displays the difficulty of fluids with different viscosities, such as mineral oil and water, to be inserted simultaneously in the micro-channel that corresponds to the spray inlet system. In this experiment, the inlet capillaries were changed to ¼" plumbing, for illustration purposes. As can be seen in the figure, after 10 s of operation, some water was in the oil inlet, causing contamination and irreproducibility in the amount of oil admitted in the system. This occurred probably by an appreciable pressure dropping in the oil inlet system, since oil has high viscosity. On such condition (figure 6B), the spray interaction area will show a high volume of oil drops that were irregular showing different volumes, and they are not dispersed in the water. On the other hand, pre-mixing can decrease the pressure drop on the spray inlet system and the interaction area will show smaller oil drops. Moreover, the pre-mixing micro-channel could be used several times with mineral oil without interaction of oil and hydrophilic areas, since oil drops were not observed on that region. Figure 6C shows drops from a spray collected on a paper sheet positioned 10 cm from the outlet. As clearly pointed out by the color difference in the collected fluid, without UVC treatment of surface of the pre-mixing system, efficient mixing cannot occur and few oil drops will be observed in the exiting fluid. When some areas of the film were exposed to UVC, an apparent completed mixed sample was obtained.

Mixing can also be useful on a gas phase. A spike of VOC-contaminated gas phase was injected inside a micro-channel with pristine or a UVC treated surface. The VOCs composition was n-hexane:2-propanol mixtures in three distinct volume ratios: 1:9; 1:1; 9:1. Because the micro-channel is long, some separation can occur [2]. Two partially distinct Gaussian shapes are seen with a pristine surface on the channel. If the film on the surface of the micro-channel was treated with UVC radiation, a clear peak is not seen, in agreement with the interpretation that it is difficult to separate n-hexane/2propanol inside the channel. In other words, a mixing of n-hexane/2propanol inside the channel was promoted.

Pre-mixing is useful on detector surfaces because it decreases the limit layer, which allows more frequent interaction between the analyte and the sensing surface. Detectors usually are planar structures, similar to the sealed cavity presented in figure 1D. The performance of this cavity was simulated with changes in hydrophilicity; figure 7 shows the main results. On a cavity with a pristine surface, the vorticity was low with almost no changes. With a partial area in the bottom surface of the structure irradiated by UVC, the vorticity profile changed while the velocity streams remained almost equal. These results indicate that the overall fluid will pass through the structure easily, although some areas will induce interaction between fluid layers, i.e. favors mixing. The experimental results show good agreement with simulation. For a liquid phase composed of glycerin (0.1 ml/min or less) and water tanned with methylene blue (5.0 ml/min), the flow was disturbed in a device with a UVC treated surface, as can be seen in figure 8A. The mixing effect can be improved using the mechanical mask presented in figure 2C. As can be seen in figure 8B, the oil drops changed the original path several times because the oleophobic/hydrophilic surface (see detail in the figure 8B) favored mixing.

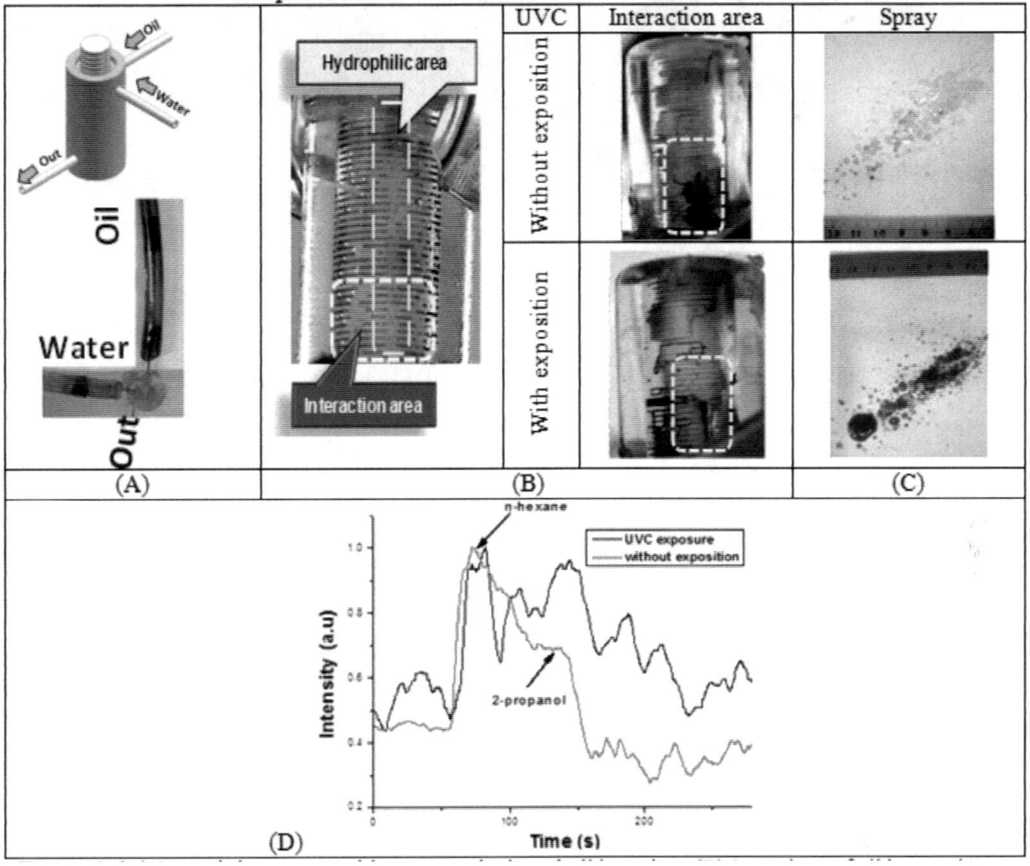

Figure 6. (A) Spray inlet system with water and mineral oil insertion. (B) Snapshots of oil interaction on a micro-channel with hydrophilic/hydrophobic areas after long use, and the spray interaction area whether pre-mixing is used or not; (C) drops formed with or without pre-mixing; (D) signal (a.u) at the outlet of a long channel UVC exposed or not.

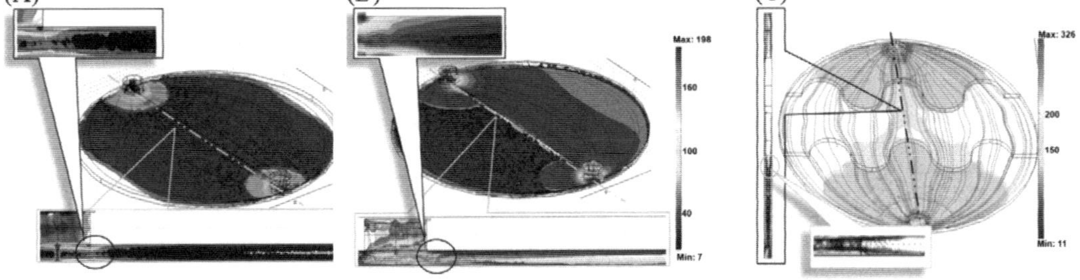

Figure 7. Simulation of the vorticity (colors, s^{-1}) and the velocity (streamline) profiles using nitrogen (5.0sccm) in a sealed cavity (A) completely homogeneous or (B) half hydrophilic or (C) exposed with mechanical mask of figure 2C.

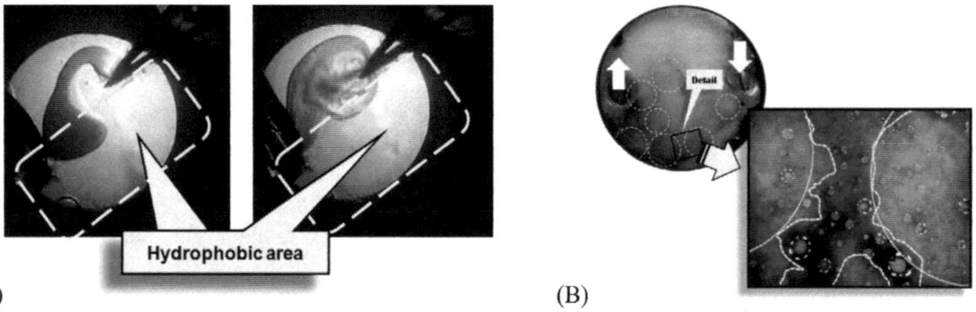

(A) (B)

Figure 8. Experimental results for a sealed cavity: (A) blue-tinted water and glycerin sample retracts towards the hydrophilic half-surface of the cavity; (B) water and oil mixes in the cavity with an alternating UVC-treated/non-treated surface (the mask for UVC-treatment is presented in figure 2C); white and yellow lines enhance the hydrophobic and tinted water areas, respectively

The films used in the previous shown experiments in the liquid phase had low sensitivity to VOCs, as quoted by QCM measurements. This is accomplished by choosing films with few carbon clusters. However, because these films are sensitive to water, they could play an important role in detecting water in the gas phase. Thus, the understanding of water-film interaction can address the design of miniaturized structures, as detectors, to improve their performance.

Figure 9 shows typical results for QCM measurements where humid air flows through a sealed cavity. In detecting humidity, TEOS films were more sensitive than TEOS/HFE films. UVC radiation degraded the sensitivity of films, whether made from TEOS or TEOS/HFE precursors, probably because fewer amounts of polymer were on the surface after the film treatment. However, UVC treatment improved water detection on TEOS films; relative humidity of 30% could be detected (Figure 9(A)).

Figure 9. Experimental results for QCM measurements of humidity on (A) TEOS or (B) TEOS/HFE films exposed or not to UVC.

4. Conclusions

This work shows advantages of using TEOS/HFE films treated under UVC radiation for the development of miniaturized structures. The performance of three different structures was tested: premixing system, a micro-sprayer and a detector for humidity. All these achievements are based on changes of surface properties, the use of a feasible procedure and the predicting and understanding of phenomena with the help of simulations. An additional benefit is that the presented films and structures are environment-friendly.

5. References

[1] Coyle C, Gandiraman R P, Gubala V, Le N C H, O'Mahony C C, Doyle C, James B, Swift P, Daniels S and Williams D E 2011, *Plasma Processes Polym.* **9** 1-28

[2] Carvalho R A M, Lima R R, Nascimento Filho A P, Silva M L P and Demarquette N R 2005 *Sens. Actuators* B **108** 1-2, 955

[3] Yang Y Q 2008 Microscale and Nanoscale Process Systems Engineering: Challenge and Progress *The Chinese Journal of Process Engineering* **8** 3 616

[4] Silva L M, Lima R R, Simões E W and da Silva M L P 2010 *Int. Rev. Chem. Eng.* **2** 134

[5] Yamasaki Y, Kariyazaki A and Morooka S 2010, *Int. J. Chem. Reactor Eng.* **8** A 140

[6] Hernandez L F, Lima R R, Pecoraro E, Rosim-Fachini E and Silva M L P 2013 *Mater. Sci. Forum* **730-732** 245, doi:10.4028/www.scientific.net/MSF.730-732.245

[7] Hernandez L F, Jesus A A, Lima R R, Fachini E R, and Silva M L P 2012 Small and Simple Devices for Increase Mixing on Detector Surfaces *14th Int. Meeting Chem. Sensors (Nuremberg) - IMCS 2012* May 20-23 AMA Science Portal doi:10.5162/IMCS2012/P2.9.2

[8] Lima R R, Hernandez L F, Pecoraro E, Rosim-Fachini E and Silva M L P 2013 *Mater. Sci. Forum* **730-732** 232 doi:10.4028/www.scientific.net/MSF.730-732.232

[9] Lima R R, Hernandez L F, Pecoraro E, Rosim-Fachini E and Silva M L P 2013 *Mater. Sci. Forum* **730-732** 289 doi: 10.4028/www.scientific.net/MSF.730-732.289

[10] Hernandez L F, Leite A R, Lima R R, Fachini E R, and Silva M L P 2012 Adsorbent Composites Used on Mixing in Miniaturized Structures *14th Int. Meeting Chem. Sens. (Nuremberg) – IMCS 2012* May 20-23 AMA Science Portal doi:10.5162/IMCS2012/P2.8.3

[11] Santos L C, Beraldo F P, Simões E W, Carvalho A T and Silva M L P 2008, *Sens. Actuators* **B 130** 310

[12] Carvalho A T, Lima R R, Silva L M, Simões E W and Silva M L P 2009 *Sens. Actuators* **B 137** 393

[13] Hernandez L F, Lima R R, Pecoraro E, Rosim-Fachini E and Silva M L P 2013 *Mater. Sci. Forum* **730-732** 185 doi:10.4028/www.scientific.net/MSF.730-732.185

[14] Dartois E. Muñoz-Caro G M, Deboffle D, Montagnac G and d'Hendecourt L 2005 *A&A* **432** 895 doi:10.1051/0004-6361:20042094

Electrospun conducting polymer nanofibers as the active material in sensors and diodes

Nicholas J. Pinto
Department of Physics and Electronics, University of Puerto Rico-Humacao, Humacao, PR 00792

E-mail: nicholas.pinto@upr.edu

Abstract. Polyaniline doped with camphorsulfonic acid (PANi-HCSA) and poly(3,4-ethylenedioxythiophene) doped with polystyrene sulfonic acid (PEDOT-PSSA) were electrospun separately to obtain individual nanofibers which were captured on Si/SiO_2 substrates and electrically characterized. The fiber resistance was recorded as a function of time in the presence of vapours of aliphatic alcohols of varying sizes. Due to the large surface to volume ratio, uniform diameter and small quantity of active material used in the construction, these sensor responses are very quick. Sensors made from individual fibers also show true saturation upon exposure to and removal of the sensing gas. A Schottky diode was also fabricated using an *n*-doped Si/SiO_2 substrate and a single PANi-HCSA fiber and tested in vacuum and in ammonia gas. The diode response was instantaneous upon exposure to ammonia with nearly complete recovery of the current upon pumping out the ammonia, thereby making it a reusable sensor with rectifying behaviour i.e. multifunctional.

1. Introduction

The rapid detection of minute traces of toxic gases in an inadequately ventilated environment is crucial to effectively reduce any widespread loss to life and property. This requires sensors capable of detecting small quantities of gas, such as those fabricated using nanofibers that possess enhanced surface to volume ratio. Organic conducting polymers (CP's) are especially suited for use as gas sensors for a variety of reasons. In particular, the oxidation/reduction states of these polymers can be reversibly tuned by exposure to basic/acidic environments, resulting in several orders of magnitude change in the conductivity [1] with no polymer degradation and that would extend the lifetime of sensors based on these polymers. Furthermore, CP's are cheap and easy to synthesize in bulk quantities, they are relatively stable under ambient conditions and can readily be processed as films or fibers.

Individual polyaniline (PANi-HCSA) and poly(3,4-ethylenedioxythiophene) (PEDOT-PSSA) nanofibers were electrospun and used separately to sense a variety of alcohols of increasing sizes with response times that are faster than sensors made from thin films of the same material [2,3]. The advantage of using a single nanofiber chemical sensor is its small size and potential for higher sensitivity. When comparing the two polymer nanofiber sensors, the PEDOT-PSSA sensors were relatively more robust and could be reused multiple times for sensing different gases. The sensors made from PANi-HCSA had reduced response after use in any gas and it was best to use a new sensor for multiple gas sensing. The ability to prepare long nanofibers of conducting polymers via electrospinning that have a large aspect ratio and an even larger surface to volume ratio, makes this technique very attractive for use in the fabrication of cheap, low power consumption and rapid response gas sensors. Finally, we fabricated Schottky diodes based on these fibers and tested them in a gaseous environment with near complete recovery [4]. This leads to the possibility of fabricating multifunctional devices making them more versatile.

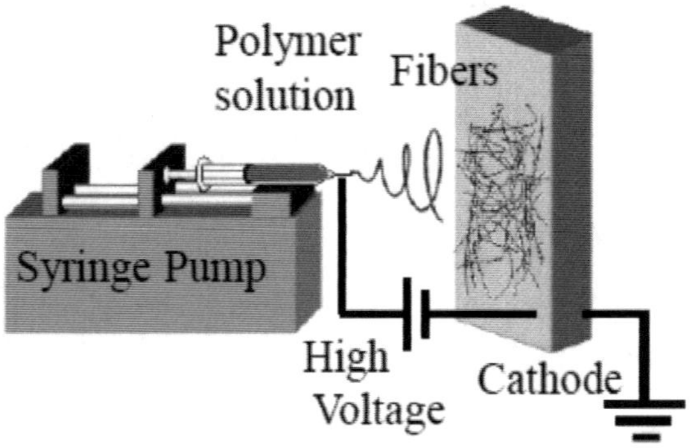

Figure 1. Basic Electrospinning Setup

2. Experimental

The basic apparatus used for the electrospinning of polymer nanofibers is shown in Figure 1. It consists of a hypodermic syringe (½ cc tuberculin syringe), a high voltage power supply (Gamma Research), a grounded cathode (Al foil) and a syringe pump (Cole Parmer).

2.1. PANi-HCSA fibers. In the electrospinning process, 100 mg of emeraldine base polyaniline was mixed with 129 mg of HCSA (solid state doping) and dissolved in 10 ml $CHCl_3$ for a period of 4 hours. The resulting deep green solution (doped polyaniline) was filtered and 3.2 mg of polyethylene oxide (PEO) having molecular weight 2,000,000 was added to the solution and stirred for an additional 2 hours. PEO was added to assist in fiber formation by acting as a plasticizer and the solution was then filtered again using a 0.45 μm PTFE syringe filter. About 0.5 ml of the solution was placed in the hypodermic syringe and the needle connected to the power supply, with the cathode (Al foil) grounded and situated about 20 cm from the tip of the needle. The flow rate was controlled to approximately one drop per 10 s. As the voltage applied to the needle increased to about 10 kV, the electric force on the polymer droplet at the end of the needle overcomes the surface tension and a polymer jet exits toward the cathode in short bursts. As the solvent evaporates in air, thin fibers of PEDOT-PSSA of varying diameters and lengths covered the substrate. Individual fibers were captured on pre-patterned doped Si/SiO_2 wafers and had diameters in the range 20-150 nm.

2.2. PEDOT-PSSA fibers. PEDOT-PSSA was purchased from Bayer Corp. (Baytron P) and used as received[5]. A 1wt% of polyethylene oxide (PEO-plasticizer) in PEDOT-PSSA was prepared and the solution used to fabricate nanofibers via electrospinning. Using the same process indicated above, individual fibers of PEDOT-PSSA were electrospun and then captured on pre-patterned doped Si/SiO_2 wafers.

An optical microscope image and an AFM image of a PANi-HCSA and a PEDOT-PSSA nanofiber are shown in Figures 2. The PANi-HCSA nanofibers were relatively long and could be manipulated with a tweezer without damage after electrospinning. The PEDOT-PSSA fibers on the other hand could not be moved after formation and were ribbon-like due to the water based solutions from which they were electrospun.

As seen in Figure 2(Right), their surface morphology was rougher than the PANi-HCSA nanofibers leading to higher surface to volume ratio. The sensors were made from individual fibers of these polymers, where the current through the fiber was monitored as a function of time in the presence of the carrier and sensing gases.

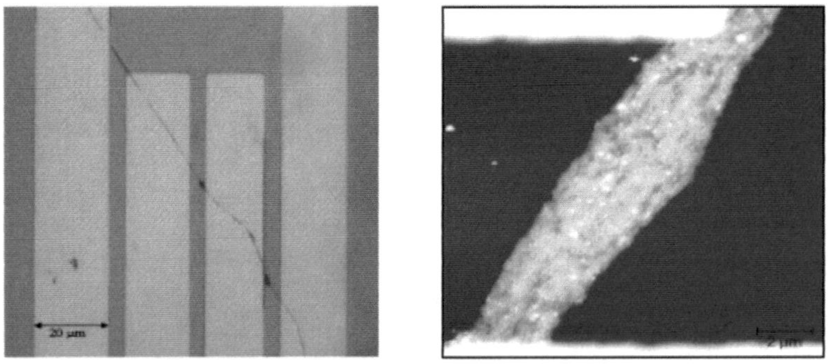

Figure 2: Left: Electrospun single PANi-HCSA nanofibers across electrodes.
Right: Electrospun PEDOT-PSSA nanober showing rough surface morphology

2.3. The Schottky diode was prepared by using an *n*-doped Si wafer (<111>, 0.1-1.0 W-cm) with a 200 nm thermally grown oxide layer and an isolated polyaniline (PANi-HCSA) nanofiber. After pre-patterning gold electrodes over the oxide via standard lithography and lift-off techniques the substrate is cleaved through the electrodes. The exposed cleaved surface has the edge of the gold electrode separated from the doped Si by the insulating oxide layer. Using the electrospinning technique reported earlier[6,7], individual, charged, dry and flexible PANi-HCSA nanofibers were deposited over the wafer edge making contacts to the gold and the doped Si and that are stable with no apparent degradation or oxidation. The resulting Schottky diode is formed along the vertical edge of the substrate at the nanofiber-doped Si interface. Such a vertical orientation may offer higher levels of integration in circuitry than that provided by in-plane horizontal structures. External electrical contacts were then made and the device current-voltage (I-V) characteristics measured via a Keithley 6517A electrometer in a vacuum of ~2 x 10^{-2} Torr. Figure 3 shows a schematic of the device and the external electrical circuit.

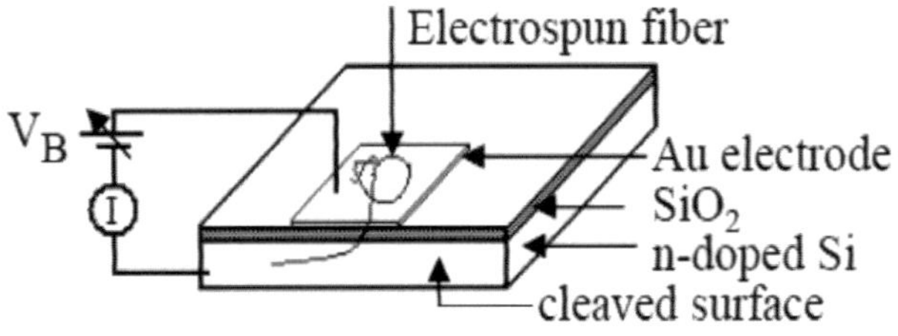

Figure 3: Schematic diagram of the Schottky diode with external electrical contacts

3. Results and Discussion

The electrospun fibers were used in aliphatic alcohol vapour sensing and also in the fabrication of a Schottky diode that was tested in the presence of ammonia vapour.

3.1 Alcohol gas sensors. Figure 4 shows the PANI-HCSA and PEDOT-PSSA nanofiber sensor response to various alcohol vapours of increasing molecule sizes.

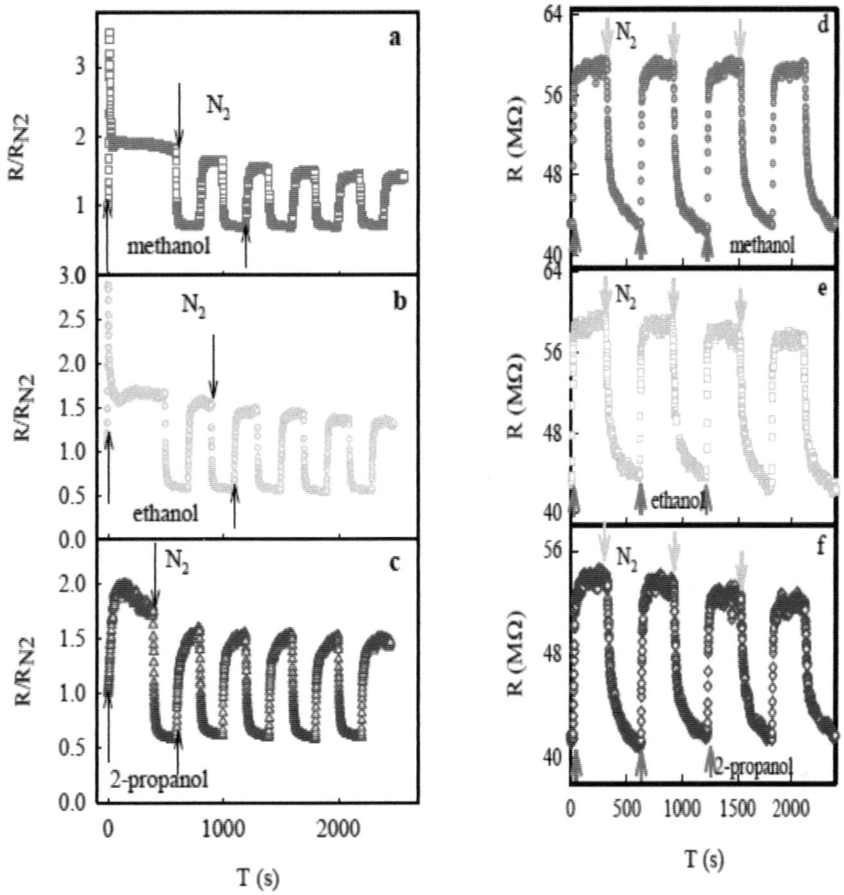

Figure 4: Time dependence of the normalized resistance of PANiHCSA(left) and PEDOT-PSSA(right) nanofibers to vapours of methanol(a), (d), ethanol(b) (e) and 2-propanol (c) (f)[2,3].

Figures 4 (a), (b) and (c) show the time dependence of the change in the normalized resistance for individual isolated electrospun PANi-HCSA doped polyaniline nanofibers as seen in Figure 2 upon exposure to vapours of methanol, ethanol and 2-propanol respectively, where R_{N2} is the resistance of the fiber under a flow of dry N_2 gas prior to the first exposure of alcohol vapour. In the case of methanol and ethanol vapours there is a rapid increase in the resistance followed by an equally rapid decrease that quickly stabilized to a relatively constant value. Such was not the case for 2-propanol as seen in Figure 4(c) where the initial increase in the resistance was slower and then the decrease was much more slower with little evidence of reaching a constant value after approximately 5 minutes. However, in all three figures we see that subsequent application of dry N_2 gas followed by the sensing alcohol vapour yield qualitatively similar results in the five complete cycles that followed viz. the flow of N_2 decreases the resistance while the flow of alcohol vapour increases the resistance. Figures 4(d) (e) and (f) show the time dependence of the change in the sensor resistance for individual isolated electrospun PEDOT-PSSA nanofibers (as seen in Figure 2) upon exposure to vapours of methanol, ethanol and 2-propanol respectively. Upon exposure to these alcohol vapours there is a rapid increase in the resistance that quickly saturated, followed by a decrease with a longer recovery time. These measurements were made on the same sensor beginning with methanol and ending with 2-propanol after which the sensor was annealed. The response times are rapid with large response amplitudes. The small fiber diameter and rough surface morphology contribute to the extremely large surface to volume ratio of the sensor. In addition, the sensor contains a small quantity of active material, hence this result is consistent and the response time is slightly faster than previous results using PEDOT-PSSA as alcohol sensors [8-11]]. The larger the alcohol molecule, the longer is the response time and smaller the amplitude response which could be related to the slower diffusion of larger molecules into the fiber. For conducting polymers where charge transport is dominated via hopping mechanisms, changes in the polymer resistance in the presence of a sensing gas occur mainly due to chemical (doping/dedoping) changes, conformational changes (tight coil vs. expanded coil) or polymer swelling (increase in inter-chain separation) [12]. When PEDOT-PSSA or PANi-HCSA is exposed to alcohol vapours, the increase in the resistance is attributed to polymer swelling [9,13] that would increase the inter-chain separation and subsequently reduce the hopping rate. This is consistent with the rapid decrease of the sensor resistance when exposed to the alcohol vapour and the slower recovery when the sensing gas is removed. Since the sensor is dried in the presence of N_2 gas prior to introducing the sensing gas, the response to the sensing gas is rapid with the alcohol vapour immediately penetrating the polymer chains causing swelling and moving them apart, then when the sensing gas is removed, the polymer chains relax more slowly to their original positions and the recovery time would depend on how effectively the sensing gas is removed. Increasing the size of the alcohol molecule increases the response time since a larger molecule would take a longer time to diffuse into the polymer. No drift in the response with time is observed which implies that the response does not affect the sensor baseline conductivity in the absence of the sensing gas and the response is therefore reversible. The sensing amplitude is also among the highest recorded for PEDOT-PSSA based alcohol sensors and must be correlated to the very large surface to volume ratio for these individual nanofibers and the high gas concentration used. Table 1 gives the response and recovery times of these sensors and the relative response amplitude changes.

Table 1. Sensor response times and amplitude response to various gases at room T

Sensing gas	PANi-HCSA nanofiber sensor			PEDOT-PSSA nanofiber sensor		
	Response time (s)	Recovery Time (s)	($\Delta R/R_{N2}$) X 100	Response time (s)	Recovery time (s)	($\Delta R/R_{N2}$) x 100
Methanol	32	20	94	8	>100	40
Ethanol	20	20	94	10	>100	40
2-Propanol	>110	>50	97	20	>100	30

3.2. Schottky diode and ammonia gas sensor. The diode prepared as mentioned in the experimental section was characterized in vacuum and in the presence of ammonia vapour. Figure 5 shows the results of this experiment in an as prepared device measured in vacuum. There is clear evidence that the curve is asymmetric and diode-like with a turn-on voltage in the range 0.4-0.6V and a rectification ratio of ~100. When ammonia gas in introduced into the chamber the current immediately gets smaller but the curve is still asymmetric. This means that the rectification ratio of the diode can be changed. Due to the large surface to volume ratio of the fiber, the change in the diode response was instantaneous upon exposure to ammonia with near compete recovery of the current upon pumping out the gas, thereby making it a reusable sensor. The recovery response can be improved by using a more efficient vacuum system. As seen from Figure 5, the diode turn on voltage is not affected by the sensing action, possibly due to the presence of surface states at the junction that control device operation. The freshly cleaved Si substrate is susceptible to containing several dangling Si bonds that would interact with the fiber upon contact thereby pinning the Fermi levels. Such pinning would make the application of this device as a sensor more reliable due to a fixed turn on voltage even in the presence of harmful gases.

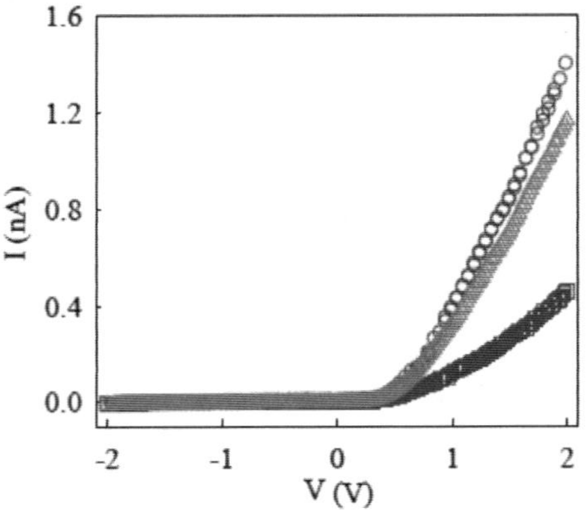

Figure 5: I-V characteristic curves for a Schottky nanodiode when used as a gas sensor: (○) -as prepared sample measured in vacuum: (□) – a few seconds after exposure to ammonia vapour: -after 5 hours of pumping out the ammonia vapour

4. Conclusions

We have successfully fabricated true conducting polymer nanofibers of PANi-HCSA and PEDOT-PSSA using the electrospinning technique and used them as sensors in the presence of various alcohol vapours and also tested a Schottky diode in vacuum and in ammonia vapour. Larger alcohol molecules have a slower response time. Nanofiber sensor responses are seen to be faster or comparable to thin film sensors and have the potential to be used in detecting small amount of gases. They can also be used in diodes that exhibit rectification and sensing capabilities and are reusable making them multifunctional.

This work was funded by NSFPREM 0934195 and NSFRUI 0965023.

5. References

[1] Chiang J C and MacDiarmid A G 1986 *Synth Metals* **13**, 193-205
[2] Pinto N J, Ramos I, Rojas R,Wang P-C and Johnson A T Jr. 2008 *Sens Actuat B* **129** 621-627
[3] Pinto N J, Rivera D, Melendez A, Ramos I, Lim J H and Johnson A T C 2011 *Sens Actuat B* **156** 849-853
[4] Pinto N J, González R, Johnson A T Jr and MacDiarmid A G 2006 *Appl Phys Lett* **89** 033505
[5] Jonas F and Morrison J T 1997 *Synth Met* **85** 1397
[6] Zhou Y X, Freitag M, Hone J, Staii C, Johnson A T, Pinto N J and MacDiarmid A G 2003 *Appl Phys Lett* **83** 3800
[7] Pinto N J, Johnson A T, MacDiarmid A G, Mueller C H, Theofylaktos N, Robinson D C F. A. Miranda F A 2003 *Appl Phys Lett* **83** 4244
[8] Sotzing G A, Briglin S M, Grubbs R H and Lewis N S 2000 *Anal Chem* **72** 3181-3190
[9] Dan Y, Cao Y, Mallouk T E, Johnson A T, and Evoy S 2007 *Sens Actuat B: Chem* **125** 55-59
[10] Jung Y S, Jung W, Tuller H L and Ross C A 2008 *Nano Lett* **8** 3776-3780
[11] Choi J, Lee J, Choi J, Jung D and Shim S E 2010 *Synth Met* **160** 1415-1421
[12] Huang J, Kaner R B, Skotheim T A and Reynolds R J(Eds.) 2006 *Handbook of Conducting Polymers*, vol. 1, 3rd ed., CRC Press, pp. 1–49 (Chapter 7) and references therein
[13] Cao Y, Kovalev A E, Xiao R, Kim J, Mayer T S and Mallouk T E 2008 *Nano Lett* **8**, 4653-4658

Effect of ethanol concentrations on few layer Schottky graphene transistors

L E G Armas[1,5], M F G Huila[2], M Pojar[1], H E M Peres[3], F J R Fernandes[3], M A Valle[3], K Araki[2], H E Toma[2], A D Santos[4] and A C Seabra[1]

[1]Laboratorio de Sistemas Integráveis do Departamento de Engenharia de sistemas Eletrônicos da Escola Politécnica da USP (LSI-PS/EPUSP), São Paulo, SP, Brazil
[2]Departamento de Química fundamental, IQ, da Universidade de São Paulo- USP, SP, Brazil
[3]Laboratorio de Microeletrônica do Departamento de Engenharia de sistemas Eletrônicos da Escola Politécnica da USP (LME-PS/EPUSP), São Paulo, SP, Brazil
[4]Departamento de Química fundamental, IQ, da Universidade de São Paulo- USP, SP, Brazil

E-mail: legarmas@gmail.com

Abstract. We have studied the effect of positive and negative back gate voltage on the Schottky barrier of few layer graphene transistors (devices) in the presence of Air, Nitrogen and different ethanol vapor concentrations. The presence of the Schottky barrier in these devices is attributed to unintentional doping of the different areas of the graphene flake and is modulated by applying a positive gate voltage, increasing the source-drain current.

1. Introduction

Interest in nanomaterials has been growing rapidly for the past several years. Particularly graphene, which is a monolayer of carbon atoms arranged in a two dimensional honeycomb structure and is the starting point for any material derived from carbon (fullerenes, nanotubes or graphite). Graphene-based rectifying devices have been investigated only scarcely due the complex processes required to achieve both *p*- and *n*-doped *semiconducting* graphene. Graphene-based *pn* junctions can be achieved by chemically doping different areas of the same graphene flake, or by appropriate electrical gating [1, 2]. However, such junctions would not be suitable for logic applications, as graphene would still retain its semimetallic character. A way around this problem would be to use graphene in conjunction with a semiconductor, in order to establish a metal-semiconductor Schottky junction [3]. Alternatively, one could reverse the role of the graphene and the semiconductor in the metal-semiconductor junction, for example, by rendering graphene semiconductor and using a traditional metal as electrode. The latter approach has been achieved and modeled by, e.g., considering semiconducting graphene nanoribbons in contact with metals [4]. A Schottky junction can also be formed by bringing semiconducting raphene oxide, obtained by either wet chemistry methods [5] or plasma oxidation [6,7], in contact with a suitable metal or even with pristine graphene itself. In a Schottky diode the forward current is due to

[5] To whom any correspondence should be addressed.

majority carriers flowing from the semiconductor to the metal; on the other hand, in a traditional semiconducting *pn* diode, minority carriers are responsible for electrical conduction.

In this work we report the effect of applying positive and negative back gate voltage on the Schottky barrier of few layer graphene (FLG) transistors in the presence of Air, N_2 and different ethanol vapour concentrations (EVCs).

2. Samples and experimental details

To fabricate the FLG devices, heavily n-doped (100) wafers with dopant concentration of $N_A = 10^{15}$ cm^3 and resistivity 0.002 – 0.005 Ω.cm were used as substrates as well as the back gate electrode with a thermally grown oxide layer of thickness of 280 nm, which acts as the dielectric, Ti/Au contacts were evaporated after optical (wider contacts) and electronic lithography (thinner contacts), and structured by the lift-off technique. After the device was fabricated, it was placed in a small chamber connected by a tube to a container of vapor mixture. Concentration of organic vapors was controlled by mixing N_2 gas and measured by gas detector. All electrical measurements were carried out by using a DC supply at atmospheric pressure.

Fig. 1(a) shows the optical image of the graphene sample (indicated by the arrows) and contacted by source (S) and drain (D) electrodes, Raman measurements were performed in three different regions of the sample, which are indicated by the numerated arrows. From the Raman spectra (Fig. 1(b)) we can see three different regions: region 1 (solid line) is a monolayer; region 2 (dashed line and middle arrow) ~ three layers and region 3 (dotted line) a bilayer graphene. The numbers in arrows are in correspondence with the numbers in the Raman spectra. On the other hand, it is well known that graphite has three intense Raman features at ~ 1580 cm^{-1} (G band), ~ 1350 cm^{-1} (D band) and ~ 2700 cm^{-1} (2D band) [8]. The G band is due to double degenerate E_{2g} mode at the Brillouin Zone center, whereas D band arises from defect mediated zone-edge (near K point) phonons, this band is present in our sample because of the small sample width which is less than 600 nm. The 2D band originates from second order double resonant Raman scattering from zone boundary, K + ΔK phonons.

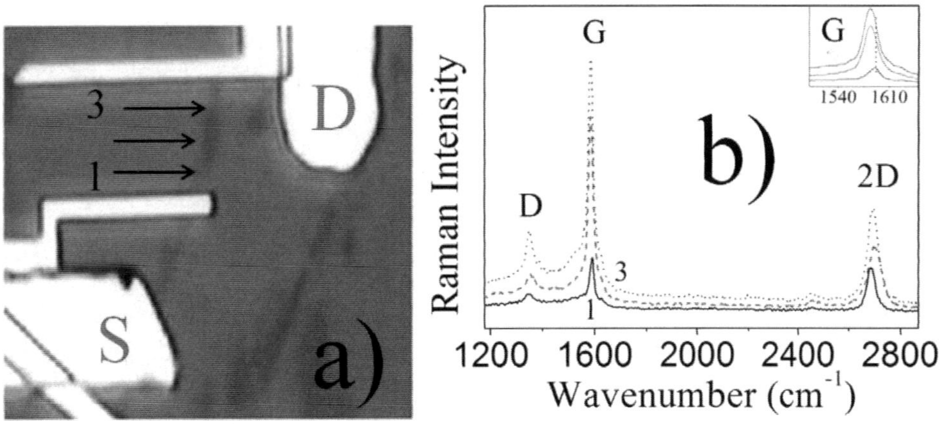

Figure 1. a) Optical image of graphene sample with three different regions, indicated by the arrows; b) Raman spectra of the three different regions of the sample, showing the presence of a monolayer (number 1), bilayer (number 3) and three layer graphene (middle arrow and curve). The inset shows a shift of the G band in the three different regions.

3. Results and discussions

Figure 2 shows the output characteristic of the studied sample, after an annealing treatment in Hydrogen (H_2)/Argon (Ar) atmosphere at 200 °C for 1 hour, in the presence of dry air and applying three different positive gate voltage. From this figure we can see the reduction of the Schottky barrier as the positive gate voltage increase. The same behaviour was observed in all the samples that were measured after annealing process. On the other hand, it is well know that the cleaning process is effective and yields significantly improved structural and electronic properties of graphene [9] without applying gate voltage, in the same way, previous studies involving dielectrophoresis (DEP) assembled reduced graphene oxide (RGO) with Au electrode have shown that mild annealing reduces the sample resistance and improves the on-current of the field effect transistors [10].

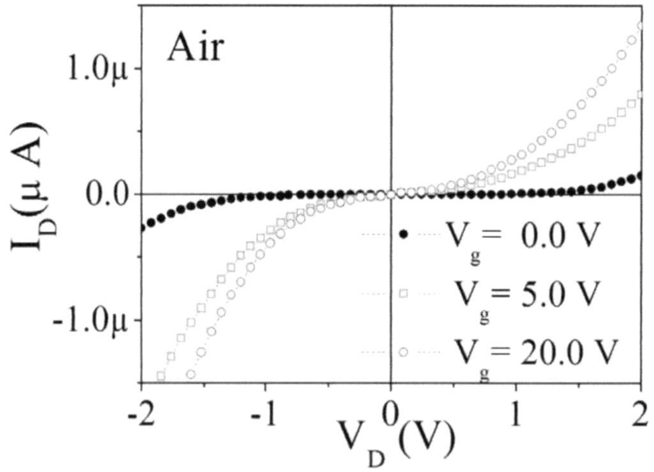

Figure 2. Source – drain current as a function of the source-drain voltage for 0.0 V, 5.0 V and 20.0 V at the presence of dry air.

With the aim of stabilizing the source - drain current, the measured were done during several cycles, in the presence of dry Air by applying different back gate voltages. After this, the measurements were done in the presence of N_2 and different ethanol vapor concentrations. The results are shown in Fig. 3. In this figure, the source – drain current (I_D) as a function of source – drain voltage (V_D) is also swept from - 2.0 V to 2.0 V, for zero (a) and negative (b) back gate voltage (V_g). Comparing Fig. 2 and Fig. 3, we observe that: the I_D current is almost zero ~ between -0.5 V and 0.5 V, the current decreased from μA (dry Air) to nA range and remains almost the same for zero and negative gate voltage at the presence of N_2, 750 ppm and 3800 ppm EVCs. The same way, we note that I_D current, in N_2 and different EVCs, is almost symmetrical for forward and reverse bias, which suggests two identical Schottky barriers (SBs) established between the FLG and the Ti electrodes. The Ti-FLG-Ti device can therefore be modeled by considering two Schottky diodes connected back to back as reported in the literature [11]. This kind of behavior was also observed in single layer graphene (SLG) samples, previously exposed to oxygen plasma treatment [11], which were named semiconducting single layer graphene (SSLG). The authors argued that after the plasma treatment a Schottky barrier (SB) arises from the difference between the work function of Chromium metal (ϕ_{Cr}= 4.5 eV) and SSLG (ϕ_{SSLG} ~ 5.0 eV). The later work function increased from 4.3 eV to 5.0 eV after the O_2 plasma treatment due to the covalent interaction between the SLG basal plane and oxygen atoms, introducing surface dipole moments. Schottky behavior was also reported in RGO samples contacted with Ti/Au electrodes [12]. However, our samples show this effect without any chemical treatment.

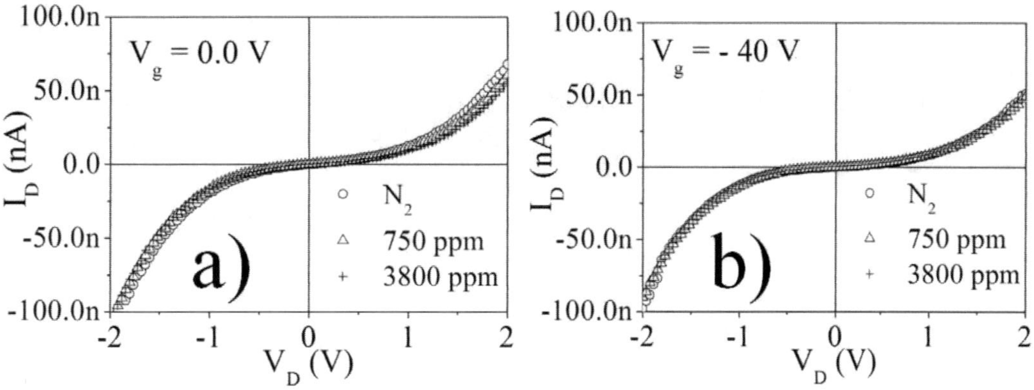

Figure 3. I_D current as a function of V_D for zero and negative gate voltage in the presence of different ethanol vapor concentrations.

The Schottky behavior of our samples could be explained in terms of unintentionally doping, which probably occurred after lithography process or during the evaporation process of thicker metal electrodes (S and D in Fig. 1). This doping probably increased the work function of FLG from ~ 4.3 eV to ~ 5.0 eV as manifested by the different peak positions of the G band (1590 cm^{-1} for region 1, 1584 cm^{-1} for region 2 and 1583 cm^{-1} for region 3) indicating a p type doping. This shifting of the G band position is similar to that reported with changes in the electronic structure and properties of graphene induced by molecular charge transfer in different samples [13]; Single Wall Carbon nanotubes (SWNTs) [14] where charge transfer to bromine and iodine is also known to cause similar shifts of the G band and on the same sample after being contacted by Cr /Au electrodes [15]. Additionally, similar results on the I_D as a function of V_D was reported on gas sensors based on FLG with different thickness on the same sample [16], in this case the authors argued that the semiconducting behavior is due to the presence of some defects in the graphene crystal structure caused by the FIB technique at the moment of fabricating their devices. Increasing of the work function in FLG sheets via chloride doping was also recently reported [17].

In order to have a better information of the device, we have measured the output characteristics, fixing the negative (- 40.0 V) and positive (+ 40.0 V) back gate voltage and varying the EVC_s as shown in Fig. 4(a) and (b). Comparing, Fig. 4(a) and (b) we can see several features on the I_D current: a) current is more symmetric for negative than for positive gate voltage, b) I_D increase when the sample is in the presence of N_2 and different EVC_s for positive gate voltage, c) positive displacement on the I_D current for high EVCs (3800 ppm) and negative gate voltage (- 40.0 V), whereas for N_2 and 750 ppm of E_{th} I_D remains almost the same. A proportional displacement of the current with V_g and V_D was also reported for three-way electrical gating characteristics of metallic Y-junction carbon nanotubes in an air atmosphere. In this system due to the interconnected nature of the Y junction, the gating is not perfect and there is a current leakage, which is manifested, in the current displacement [18]. However, in the studied device the gate leakage current is in the range of pA, because of, the FLG channel and the gate electrode (substrate) are separated by the 280 nm SiO_2 dielectric, which do not allow passing high gate current from substrate to the graphene channel.

Increasing on the I_D current, as the positive applied voltage is increased, could be attributed to a reduction of the Schottky barrier height which is unaffected by the presence of Air, as shown in Fig. 2, N_2 and different ethanol vapor concentrations as shown in Fig. 4(b). On the other hand, modulation of the Schottky barrier by applying positive gate voltage was reported in carbon nanotube field effect transistors-based NH_3 gas sensors [19].

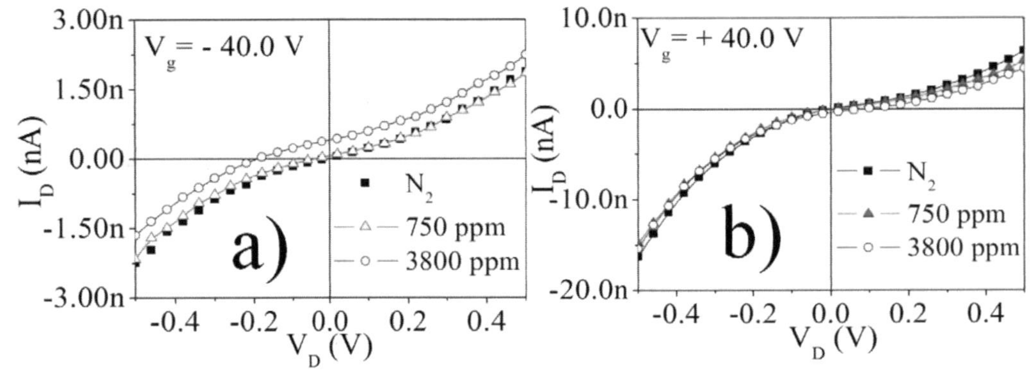

Figure 4. Source – drain current (I_D) as a function of the source-drain voltage (V_D) at the presence of N_2 and 3800 ppm of EVC, varying the applied back gate voltage.

4. Conclusions

In conclusion, we have reported that the Schottky barrier in FLG transistors could be originated from a p-type doping of different areas of FLG sample and could be modulated applying positive and negative back gate voltage in the presence of Air, N_2 and different ethanol vapor concentrations.

5. References

[1] Williams J R, Dicarlo L and Marcus C M 2007 Science **306** 666
[2] Lohmann T, Von Klitzing K, and Smet J H 2009 Nano. Lett. **9** 1973
[3] Li X, Zhu H, Wang K, Cao A, Wei J, Li C, Yia Y, Li Z, Li X and Wud D 2010 Adv. Matt. **22** 2743
[4] Jena D, Fang T, Zhang Q and Zhing H L 2008 Appl. Phys. Lett. **93** 112106
[5] Lu Z T, Vora P M, Mele E J, Johnson A T C and Kikkawa J M 2009 App. Phys. Lett. **9** 111909
[6] Gokus T, Nair R R, Bonetti A, Bohmler M, Lombardo A, Novoselov K S, Geim A K, Ferrari A C and Hartschuh 2009 ACS Nano. **3** 3963
[7] Nourbakhsh A, Cantoro M, Vosch T, Pourtois G, Clemente F, Van der Vem M H, Hofkens J, Heyns M M, De Gendt S and Sels B F 2010 Nanotechnology **21** 435203
[8] Das A, Chakraborty A and sood A K 2008 Bull Mater Sci. **31** 579
[9] Dan Y, Lu Y, Kibert N J, Luo Z and Charlie Jonhson A T 2009 Nano. Lett. **9** 1472
[10] Dan J, Chunder A, Zhai L and Hhondaker S I 2010 Nanotechnology **21** 165202
[11] Nourbakhsh A, Cantoro M, Hadipour A, Vosch T, Van der Veen M H, Heyns M M, Sels B F and De Gendt S 2010 Appl. Phys. Lett. **97** 163101
[12] Islam M R, Joung D and Hhondaker S I 2011 New J. Phys. **13** 035021
[13] Das B, Voggu R, Rout C S and Rao C N R 2008 Chem. Commun. **41** 5155
[14] Voggu R, Rout C S, Franklin A D, Fisher T S and Rao C N R 2008 J. Phys. Chem. C **112** 13053
[15] Casiraghi S, Pisana S, Novoselov K S, Geim A K and Ferrai A C 2007 Appl. Phys. Lett. **91** 233108
[16] Massera E, La Ferrara V, Miglieta M, Polichetti T, Nasti I, Di Francia I 2011 Chemistry Today **29** 39
[17] Kwon K C, Choi H S and Kim S Y 2012 Adv. Func. Mat. **22** 4724
[18] Park J, Daraio C, Jin S, Bandaru P R, Gaillard J and Rao A M 2006 App. Phys. Lett. **88** 243113
[19] Peng N, Shang Q, Lee Y C, Tan O K and Marzari N 2008 Sens. Actuat. B **132**

Sensors based in suspended gate field effect transistors for pH measurements

B da Silva Rodrigues[1,2,3], A L Siarkowski[1], O De Sagazan[2], S Crand[2] T Mohammed-Brahim[2] and N I Morimoto[1]
[1]LSI, Universidade de São Paulo, Av. Prof. Luciano Gualberto, trav. 3 n°158, São Paulo, Brasil
[2]GM-IETR UMR-CNRS 6164, Universite RENNES I, 35042 Rennes Cedex, France

Email: bruno@lsi.usp.br

Abstract. This work presents a new pH sensor based on the architecture SGFET (Suspended Gate Field Effect Transistor). The pH variations are obtained using the threshold voltage and the transfer characteristics shift of the SGFET. To evaluate the accuracy and reliability of the pH sensor, pH variations were measured directly in the solution using both the SGFET and a commercial pH-meter.

1. Introduction

The usual glass pH-meter are often being replaced by Ion Sensitive Field Effect Transistors (ISFET's). Since Bergveld [1] first employed the field effect transistor in neuro-physiological measurements in 1970, ISFETs have been developed into a new type of chemical sensing electrode. Many theoretical and experimental studies have been published describing the behavior of this chemical-sensing electronic device [2,3].

The sensitivity of ISFETs can be increased by using a suspended gate at a sub-micrometric distance from the channel of the transistor [4]. This sub-micrometric distance allows the field effect to increase the sensitivity by more than 7 times the Nernstian sensitivity [5]. This new technology is called Suspended Gate Field Effect Transistor (SGFET) and was originally developed by Janata [6]. Other variations of the structure followed after the SGFET, such as Hybrid SGFET [7] and Floating Gate FET [8]. These first structures were developed only with the goal of fixing the usual reference electrode of ISFETs because the suspended gate was too far from the transistor channel to benefit from the field effect.

In this paper, technological improvements of the ISFETs structure are studied with the goal of obtaining practical, simple to use and highly sensitive pH-meters. This work presents different ways of polarizing SGFET's and the relative pH sensitivity of each method. Using the results of polarization tests the architecture of an array-based micro-system has been developed and tested.

1.1. SGFET process and electrical principle

Figure 1 shows a scheme of a P-type SGFET just before removing the Undoped Sicila Glass (USG) sacrificial layer. Low-doped N type silicon wafers with <100> orientation were used as

[3] To whom any correspondence should be addressed.

Published under licence by IOP Publishing Ltd

substrates. The drain and source wells come from boron nitride diffusion at 1100°C with a 1μm SiO_2 as protective layer. The channel is then oxidized at 1100°C so that the SiO_2 thickness is kept at 70 nm. After that, the channel oxide is protected by a 50 nm Si_3N_4 LPCVD (Low Pressure Chemical Vapour Deposition) layer deposited at 600°C. On the silicon nitride layer, a 500 nm USG sacrificial layer is deposited by Plasma-Enhanced Chemical Vapor Deposition (PECVD) at 350°C, followed by a 50 nm silicon nitride layer which acts as isolation layer for the bottom of the suspended gate.

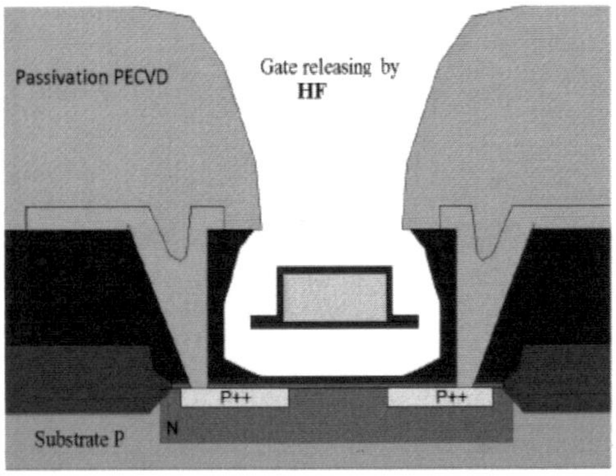

Figure 1. Scheme of the structure of P type SGFET before releasing of the USG sacrificial layer.

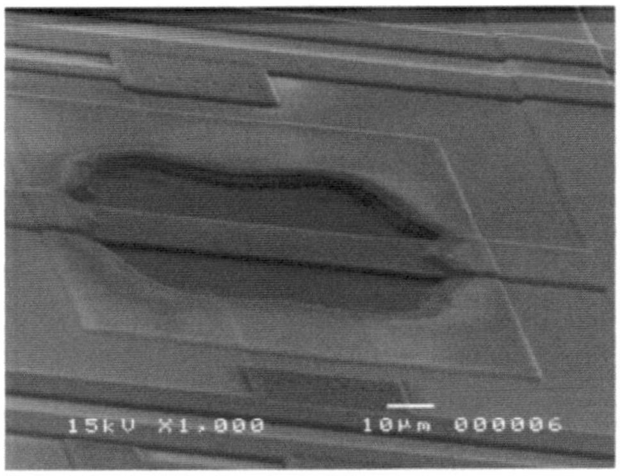

Figure 2. SEM micrograph of the SGFET structure after releasing of the sacrificial layer.

Gate and drain source contacts are then obtained by adding a 500 nm thick P-type polysilicon layer amorphously deposited at 550°C and crystallized during 12h at 600°C. A last 50 nm of a nitride layer protects the top of the gate. Vias are opened in this silicon nitride layer to make the contacts with Al path. All surface metal layers are protected with oxide layers such as BPSG (Boro Phosphorous Silica Glass) and nitride (PECVD) to avoid direct electrical leakage during the use in a liquid environment. The last step of the process is the release of the gate by HF etching of the USG sacrificial layer. A Scanning Electron Microscope view of the final structure is shown in Figure 2.

For this architecture, the principle of ph-meter measurement is based in the concentration of charges in the gap between channel and gate. The concentration of H^+ or OH^- in the gap influences the threshold voltage (V_T) of the SGFET as shown in equation 1,

$$V_T = \phi_{MS} + 2\phi_F + \frac{Q_{ss}}{C_{ox}} - \frac{1}{Ce_{ox}}\int_0^{e_{ox}} x_\rho(x)dx \qquad (1)$$

Where ϕ_F is the Fermi potential in the bulk of the n-type semiconductor, ϕ_{MS} is the difference in the work function between the metal and the semiconductor, Q_{ss} is the density of charges in the effective gate oxide, C_{ox} is the total capacitance between the gate and the semiconductor, e_{ox} is the thickness of the insulation and ρ is the charge in the insulator at a distance x from the gate.

2. pH measurements

In a p-type SGFET, the absolute value of V_T (threshold voltage) increases as the pH decreases. In the experiments presented in this paper, adding small amounts of the basic solution varied the pH of the acid solution. After waiting a few seconds, the transfer characteristics of the SGFETs were measured. Figure 3 shows the transfer characteristics and the trans-conductance of a P-type ISFET for different pH values.

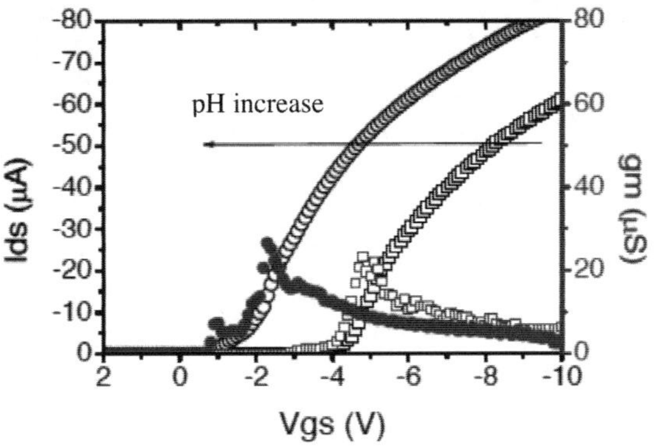

Figure 3. Transfer characteristics $I_{ds}=f(V_{gs})$ and trans-conductance (g_m) of P-type SGFET in different pH solutions with V_{ds}= -0.3 V.

The trans-conductance was calculated using the transfer characteristics. The sensitivity of the SGFET sensors was calculated using the Vgs values for the maximum trans-conductance. A sensitivity of approximately 630mV/pH was obtained. This value is much higher than those reported in literature for usual pH-meters. Actually, the sensitivity of usual pH-meters is limited by the theoretical Nernstian response of 59 mV/pH at 20°C.

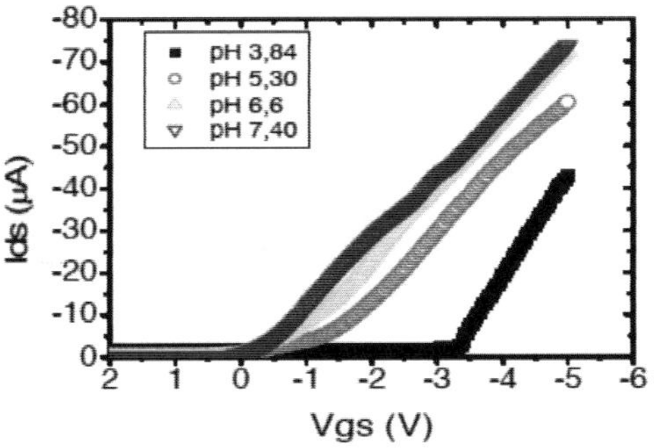

Figure 4. Transfer characteristics $I_{ds}=f(V_{gs})$ P-type of P-type SGFET in different pH solutions with V_{ds}= -0.3 V.

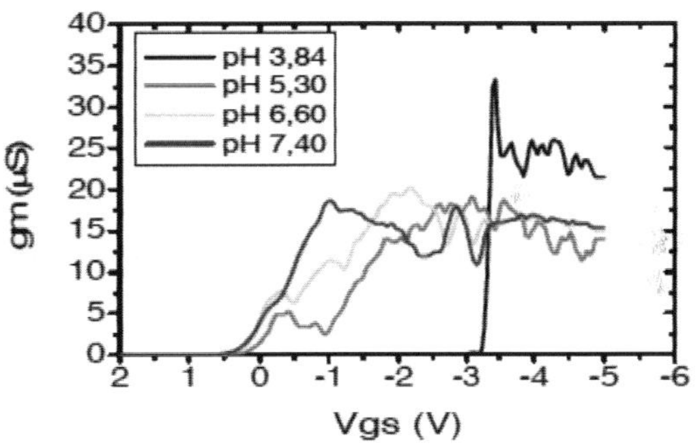

Figure 5. Gm=f(V_{gs}) trans-conductance SGFET for different solutions and V_{ds}= -0.3 V.

Figure 4 shows the transfer characteristics for solutions with different pH values. Figure 5 shows the trans-conductance obtained from the transfer characteristics. A sensitivity of 650 mV/pH is obtained from the linear relationship between Vgs and Ph, shown in figure 6. This result demonstrates the high sensitivity and reliability of SGFET's sensors.

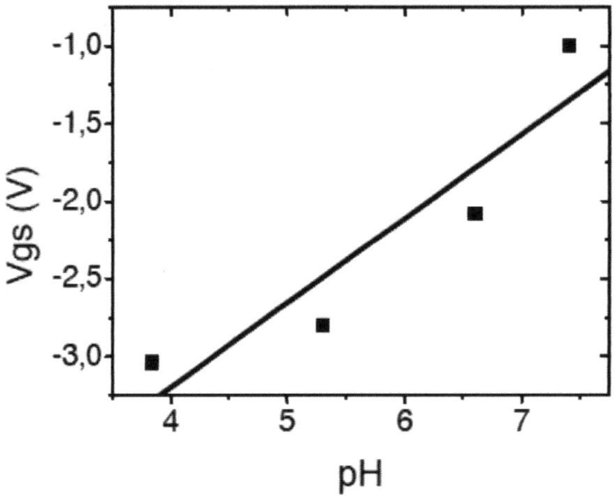

Figure 6. Gate voltage V_{gs} extracted from the maximum trans-conductance for different pH solutions.

3. Conclusion

An SGFET sensor that functions as a fully electronic pH-meter has been developed using standard microelectronics technology. The detection principle is based on the field effect mechanism of a MOSFET. The system reacts to the shift of the gap capacitance due to pH variation. The measurements demonstrated the simplicity of the method used to bias the sensor and to analyze the output signal. It has been shown that a pH variation could be detected with a good sensitivity of approximately 650 mV/pH. Finally, the method presented has good reliability and sensitivity.

Acknowledgments

The authors would like to thank Christian Le Mouellic, Patrice Lancelot and Joseph Tregret from MHS (Nantes-France) for their help and advice in manufacturing, NAMITEC, FAPESP, CAPES and CNPq the financial support.

4. References

[1] Bergveld P, 1970 Development of an ion sensitive solid-state device for neurophysiological measurements *IEEE Trans. Biomed. Eng.***17** 70-71

[2] Clifford D Fung, Peter W Cheung and Wen H. Ko 1986 A generalized theory of an electrolyte insulator semiconductor field effect transistor *IEEE Trans. Electron Devices* **33** (1) 8-18

[3] Bergveld P 2003 Thirty years of ISFETOLOGY What happened in the past 30 years and may happen in the next 30 years *Sensors and Actuators B* **88** 1-20

[4] Bendriaa F, Le Bihan F, Salaün A C, Mohammed-Brahim T and O. Bonnaud 2005 SPIE's *International Symposium on Microtechnologies for the New Millennium* Sevilla Spain

[5] Rodrigues B S De Sagazan O; Crand S, Mohamed Brahim T. and Morimoto N I 2009 *The Electrochemical Society* **23** 203-209

[6] Janata J 1983 *Sensors and Actuators* **4** 255-265

[7] Flietner B, Doll T, Lechner J, Leu M, Eisele I 1994 *Sensors and Actuators B* **18-19** 632-36.

[8] Burgmair M, Frerichs H P, Zimmer M, Lehmann M, Eisele I 2003 *Sensors and Actuators B* **95** 183-188

Journal of Physics: Conference Series **421** (2013) 012007 doi:10.1088/1742-6596/421/1/012007

Electrochemical characterization of poly-(3,4 propylene-dioxythiophene) pseudo-capacitor

H Sahoo, JW Park, RA Cardona and J Santiago-Avilés[1]
Electrical and Systems Engineering Department
University of Pennsylvania, Philadelphia, PA 19104

E-mail: santiago@seas.upenn.edu

Abstract. Every day use of appliances relies mostly in lithium-ion batteries to satisfy their energy requirements. However, the materials utilized and their lower power densities limit these batteries' desirability. An alternative to batteries is the supercapacitors, which are capable of storing energy in the electrical double layer (EDL) formed between the electrode material and the electrolyte. To reduce the gap in terms on energy and power density between batteries and EDL supercapacitor, pseudocapacitors has been used. In pseudo-capacitors a material that is capable of storing faradaic charge, such as metal oxides and conducting polymers, is deposited in the electrode surface, but its charge / discharge behavior approximate that of the EDL supercapacitor. Therefore, energy density is gained even though the faradaic nature of the process makes its power density decreases. In this paper, we use the conducting polymer, poly-(3,4 propylene-dioxythiophene) for the assembling of a pseudo-capacitor. We present the electrochemical characterization of the devices as a function of the amount of material accumulated in the platinum current collector, in terms of the capacitance, energy, and power density.

1. Introduction

Energy consumption worldwide increased from 495 kWh in 2007 with a projected amount of 739 kWh by 2035 [1]. From this total, countries belonging to the OECD (Organization for Economic Co-operation and Development) currently consumed 50 % (~ 230 kWh). However, when this number is estimated to remain constant for OECD's countries by 2035, the emerging economies will be accountable for the near-total increment in global energy consumption. This panorama calls for actions to improve the use and management of electrical energy resources.

In the USA, nearly 41 % of the energy produced goes into electric power, form which nearly a 50 % it used to provide electricity to residences and for commercial purposes [2]. Regardless of the challenges to obtain energy from renewable sources, a prominent one is to determine how to effectively storage the unused electric energy during low load periods. Energy storage devices (ESD) can present an alternative for the managing of electric energy. Also, ESD are important for the miniaturization of technologies such as those in demand for household electrical appliances, smart phones, personal computers, biomedical applications (peacemakers), and electric vehicles.

[1] To whom any correspondence should be addressed.

Supercapacitors are just one of the approaches to storage energy, among others such as the Li-ion batteries, and fuel cells. Super-capacitors rely on the formation of the EDL between the material deposited in the electrode (regularly carbon based materials) and the electrolyte [3]. Different from batteries, in supercapacitors no chemical reaction is going on, therefore the cyclability of these ESD is higher as well as its power density. However, energy density is lower than that of the batteries which limits it uses in some applications. In pseudo-capacitor an electroactive material is placed in the electrode, in a configuration that permits the formation of the double-layer phenomena [3]. Pseudo-capacitors have appeared as alternative to reduce the gap between purely EDL super-capacitor and Li-ion batteries, by combining chemical phenomena and surface physics phenomena.

In this study we present the use of electro-active polymer poly-(3,4 propylene-dioxythiophene (p-ProDOT) as material for the construction of a pseudo-capacitor. There have been previous studies using p-ProDOT as conducting polymer for pseudo capacitors [4,5] however, none of these studies explore in depth the dependency of material accumulated in the electrode and capacitance. In the present studies we will discuss how the amount of material deposited over electrode's surface affect the characteristics of the device including, capacitance, energy and power density.

2. Materials and Instrumentation

2.1 Materials

3,4 propylenedioxythiophene (ProDOT), 99%, KCl and Propylene carbonate (PC), 99.7% anhydrous were obtained from Aldrich. Tetrabutyl ammonium perchlorate (TBAP), 99% was obtained from Fluka Chemical, Potassium ferricyanide, ACS grade, was from MP Biobedicals, LLC, diamond paste (No. 40-6244), Metadi fluid (No. 40-6032) were from Buehler Co., Water, D.I.U.F. was obtained from Fisher, an separator membrane from GorTEX ®.

2.2 Instrumentation

A Bipotentiostat (model AFCBP1) by the PINE Instrument Company and a glove box from the LC Technology Solution Inc. have been used for all of the experiments. In addition, a potentiostat / Galvanostat (model 273 A) by the EG&G Princeton Applied Research was used for the platinum electrode surface area determination.

3. Methods

3.1 Electrode area calculation

A three-electrode cell, with platinum electrodes as the working electrode, Ag/AgCl (NaCl 3 M) and Nichrome wire as the reference and auxiliary electrodes respectively was use. All potentials are reported against the Ag/AgCl (NaCl 3 M), electrode. The Pt electrodes surface was polished with diamond paste/metadi fluid in a microcloth and rinsed with water. Subsequently, the electrode was cycled in 1 M H_2SO_4 solution from (-200 to 1200) mV at 100 mV/sec for 50 cycles, rinsed with water and air dried [6].

To calculate the platinum electrodes area, the standard procedure [7] using potassium ferricayanide was employed. Chronoamperometry for a 1 x10^{-3} M $K_3Fe(CN)_6$/ 0.1 M KCl solution was done. The resulting chronoamperogram is converted to the corresponding Anson Plot (charge vs time$^{1/2}$) and using the slope of the forward step and the derived Cottrell's equation (Equation 1, where m is the slope, n is the number of electrons transfer per mol, F is Faraday constant (96, 485.30 C (mol e-), C* the bulk concentration of $K_3Fe(CN)_6$, and D is the diffusion coefficient of $K_3Fe(CN)_6$ (7.63 x 10^{-6} cm^2/s)) [8]. The above relation allows us to calculate the electrode area. This procedure was done three times for each electrode and the average area reported.

$$A = \frac{m\pi^{1/2}}{2nFC\cdot D^{1/2}} \tag{1}$$

3.2 Electro-polymerization and doping

Using a three-electrode cell as described above, cyclic voltammetry was performed using a 5 mM of ProDOT/0.1M TBAP/PC solution. The voltage window used was from -1000 mV to 1500 mV at a scan rate of 100 mV/s. The number of cycles went from 10- 50 in increments of 10. The same procedure was performed for two electrodes. The Coulombic efficiency was measured to assure that the electrodes do not differ from each other by more than a 30%. Once the electrochemical polymerization process was completed, the surface of the electrode was washed with a few drops of 0.1 M TBAP/PC to remove the excess un-reacted monomer that may be present on the surface. The electro-polymerization and doping chemical reactions are showed in figure 1.

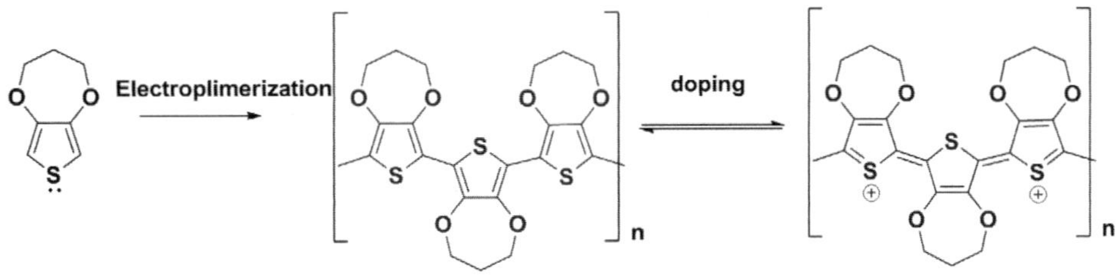

3,4- propylenedioxythiophene

poly-3,4- propylenedioxythiophene

Figure 1. Electro-polymerization and doping reaction for poly-3,4-propylenedioxythiophene

Doping of the polymer is carried out in a similar three-electrode system with 0.1M TBAP/PC as the electrolyte. Out of one of the electrodes polymerized in the previous step, an oxidative reaction is utilized to create a positive charge. The setup is subjected to analog voltammetry sweep at a scan rate of 100 mV/s going from 0 to 700 mV. A complete oxidation of the polymer is accompanied by a color change from light shade of purple to a navy blue. This was considered as the stopping point for the oxidation process. The other electrode is neutralized by analog voltammetry sweep at a scan rate of 100 mV/s using a voltage window of -500 mV to 500 mV for 100 cycles, to ensure that there are no stray charges left on the polymer.

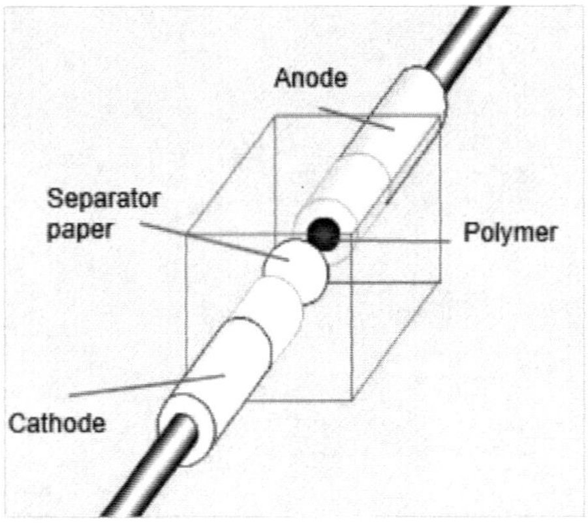

Figure 2. Pseudo-capacitor device configuration.

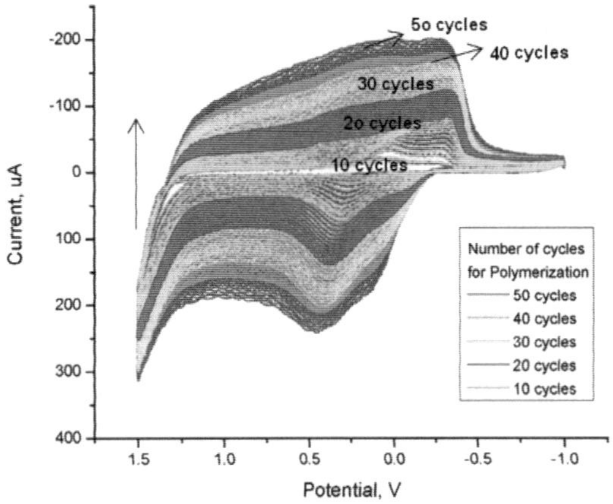

Figure 3. Cyclic voltammograms for the electro-polymerization of 5 mM 3,4-propylenedioxythiophene at 100 mV/s in 0.1 M TBAP/ PC at different polymerization cycles.

3.3 Device Test
The two electrodes are then put together in an assembly shown in figure 2. A few drops of 0.1 M of TBAP are used as electrolyte and a separator paper is used in between the electrodes. For the device test cyclic voltammetry (4 cycles) at different working windows (1 V, 1.5 V, and 2 V) is done at different scan rates (25 mV/s-10,000 mv/s). From the results the capacitance, energy density, power density and device life was determined.

4. Results

4.1 Electropolymerization and doping
Calculated Pt electrodes' area ranged from (0.018-0.020) cm^2. Figure 3 shows the cyclic voltammograms as function of polymerization cycles obtained for the electro-polymerization of poly-3,4-propylenedioxythiophene. The trace crossing of the reverse scan, associated with the initiation of the nucleation process that occur as initial products are formed (dimers, oligomers, and bipolarons), is observed in the first cycle, as reported previously for conducting polymers [8]. The amount of material deposited in the electrode was calculated using the equation for quantitative estimate of the surface coverage for adsorbed species (Γ=Q/(nFA)) [9,10] where Q is the charge in coulombs (obtained from the integration of the voltammetric wave), A is the area of the electrode in cm^2, F is the Faradays' constant (96,487 coulombs/ mol), and n is the number of electrons per mol of reaction that, for p-PRODOT, yields 2.25 electron/monomer [5]. By simply multiplying by the monomer molecular weight, the mass of material deposited in the surface can be calculated. Table 1 shows the mass in micro grams of the material deposited over the electrode as a function of polymerization cycles. As cycling is increased the amount of material deposited in the electrode increases, as expected. However, it can be appreciate from the table that the tendency is to reach a plateau as the numbers of cycles approach 50 cycles.

We attribute this result to a diffusion limitation, as more material is deposited, few monomer molecules from the solution can accesses the electrode surface and therefore the current (which have a direct dependence with the square root of diffusion coefficient [9,10], is reduce). This can be observed from the voltammograms in figure 2. Going from 30 to 50 polymerization cycles, the increment in current is considerably reduced. Also, a close look to the oxidation wave show a shift to more positive potentials as polymerization cycles are increased, which denotes that the oxidation process becomes more difficult as more material is grown in the surface [11].

Table 1. Mass of the material accumulated as a function of polymerization cycles

Number of Cycles	Mass, µg
20	0.68
40	1.19
60	1.66
80	1.93
100	2.09

The doping time dependence with number of cycles, and therefore amount of material deposited was also determined. It can be observed in figure 4, that as the amount of cycles are increased, the time for doping increases exponentially. As more material is deposited in the electrode, the accesses of the electrolyte to all the oxidation sites become harder and therefore, the oxidation time increases.

4.2 Device test
The cyclic voltammograms for the device test of the p-ProDOT pseudocapacitor using a one-volt window at different scan rates are shown in figure 5. The voltammograms showed the expected

square-shape for a super capacitor. As working window is opened (results not showed) the square shape is lost at lower scan rates.

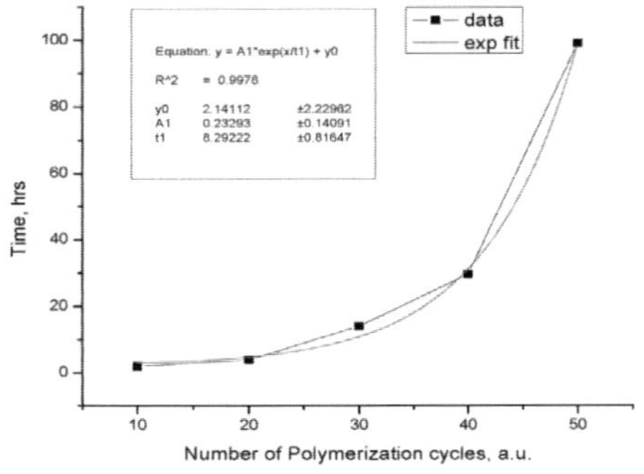

Figure 4. Oxidation time dependence with amount of polymer deposited (polymerization cycles).

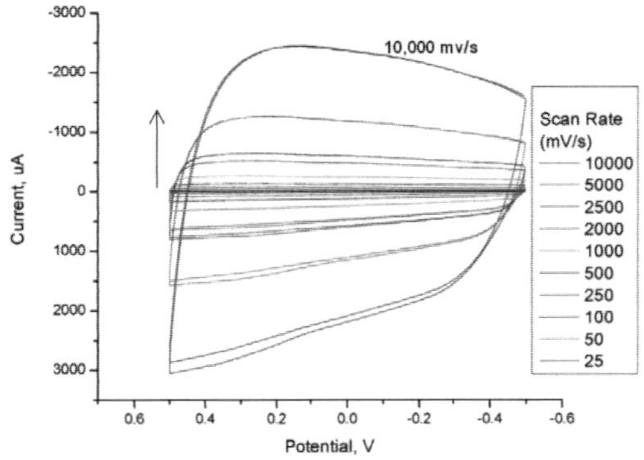

Figure 5. Cyclic voltammetry for device test at different scan rates

It can also be appreciated as expected, an increment in current as the scan rate is increased. The capacitance as a function of scan rate and frequency, and dependence on polymerization cycles for a working window of 2 volts is show in figure 6. The capacitance is determined from the charge of capacity ($C_{1/2}$) and the device voltage (V_d) and corrected for the amount of material in the electrodes. The capacitance profiles for all the devices show the expected decay in capacitance as scan rate is increased.

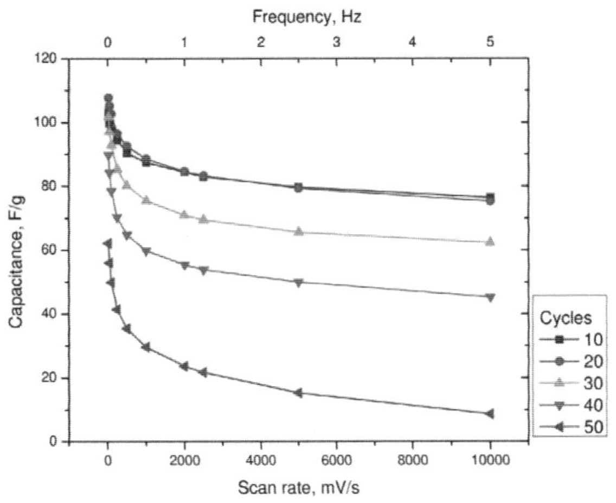

Figure 6. Capacitance as a function of frequency and scan rate for 2 volts working window for different polymerization cycles

The finer details can be seen more clearly form the plot of capacitance with the log of scan rate (figure 7A). This transformed the results plot into a linear decay, and from the slope for these lines it can be observed that as the polymerization cycles are increased, the slope increases (2.29 $\pm$ 0.59 to 4.46 $\pm$ 1.20), meaning that the discharge for the higher polymerization device is more dramatic as frequency increases. The capacitance for 10 and 20 cycles of polymerization is similar, with a slightly difference of 4 F/g (103.3 F/g vs 107.8 F/g) at lower frequencies with a lost in capacitance of approximately 30 F/g from scanning all the frequencies ranges. For 30 cycles we see a difference ranging from 101.8 F/g to 62. F/g. The most dramatic is for 40 and 50 cycles, that even at low frequencies there is a difference in capacitance of up to 40 F/g in comparison with the other devices, and for the 50 cycles we see that at the highest frequency the capacitance abruptly decreases (8.7 F/g). Similar tendencies are observed in figure 7B and 7C for power and energy respectively. These observation can be explain based in diffusion limitations, even when as more material is deposited in the electrode and therefore more possible charges, accessing them is more difficult. Since capacitance is a surface phenomenon between the electrolyte and the electrode material, it is essential for the electrolyte to be in contact with the electrode. Sulaiman and Kataky [8] reported that the morphology of electro-polymerized p-ProDOT, different form PEDOT that posses a rough dense morphology, have a very porous surface with micro-sized holes. Based on this observation we conclude that as more material is accumulated in the electrode surface, the accesses of the electrolyte to these holes during the experimental time frame becomes more difficult, therefore reducing the device overall capacitance.

The device cyclability was also calculated for 10,000 cycles, for 20, 30, and 40 polymerization cycles devices. At a scan rate of 100 mV/s and a working window of 2 V (figure 7D), the results show that after 5 % of the initial current lost or less, the device keeps stable during the time scale of the experiment for all the tested parameters.

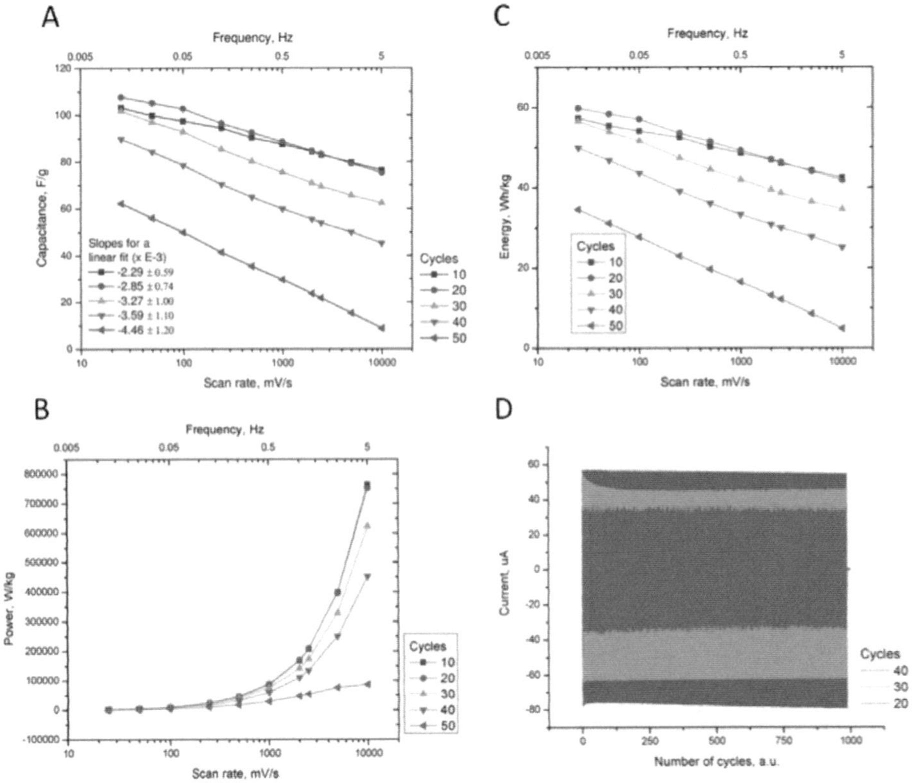

Figure 7. Device Characteristics for 2 Volts working window A) Capacitance as a function of frequency and log scan rate B) Power as a function of frequency and log scan rate C) Energy as a function of frequency and log scan rate D) Stability of the device for 1000 cycles as a function of the polymerization cycles.

5. Conclusions

The dependence of p-ProDOT pseudo capacitor characteristics upon the amount of material deposited in the electrode has been investigated. The results show an increment in capacitance as function of frequency after 30 polymerization cycles. The slope of the device for 50 polymerization cycles experiences a two folds increment in comparison to 10 and 20 polymerization cycles. This behavior has been attributed to a diffusion limitation of the electrolyte to access all the sites in the polymer as the amount of material is increase.

Acknowledgements

We would like to acknowledge support from the following sources: The Penn's Nano-Bio Interface Center (NBIC) through the NSF sponsored grant NSEC DM R08-32802. R. Cardona would like to acknowledge the University of Pennsylvania Provost Office and the Laboratory for the Research on the Structure of Matter (LRSM) through the NSF grant DMR11-20901 for her support as a postdoctoral fellow.

6. References

[1] EIA *Annual Energy Review 2009. http://205.254.135.24/totalenergy/data/annual/pdf/aer.pdf accessed 5/7/2012*

[2] Ibrahim H, Ilinca A and Perron J 2008 *Renewable & Sustainable Energy Reviews* **12** 1221-1250

[3] Abruña H D, Kiya Y and Henderson J C 2008 *Physics Today* 43-47

[4] Rosario-Canales M R, Deria P, Therien MJ and Santiago-Avilés J J 2012 *Appl. Mater. Interfaces* **4**102–109

[5] Bobacka J, Lewenstam A, Ivaska A *J 2000 Electranal. Chem.* **489** 17-27

[6] Doña-Rodríguez J M, Herrera-Melián J A, Pérez-Peña 2000 *J.J. Chem. Educ.* **77** (9) 1195-1197

[7] Kissinger P T, Heineman W R 1996 *Laboratory Techniques in Electroanalytical Chemistry* (2nd ed. Marcel Dekker, Inc. New York, USA) 11-49; 141-159; 403-439

[8] Sulaiman Y and Kataky R *J 2011 Electrochem. Soc.* **2011 159** (2) F1-F9

[9] Kissinger P T and Heineman W R 1983 Cyclic Voltammetry.J. Chem. Educ. **60** (9) 702-706

[10] Abruña H D 1988 Coordination Chemistry in Two Dimensions: Chemically Modified Electrodes *Coord. Chem. Rev.* **86** 135-189

[11] Osaka T, Naoi K, Ogano S and Nakamura S *J. Electrochem. Soc.: Electrochemical Science and Technology.* **134** (9) 2096-2102

Synthesis, characterization and use of Ru-Fc intercalation complex as an electrochemical label for the detection of pathogen-DNA

M Díaz-Serrano[1], A Rosado[1], D Santana[2], E Z Vega[2] and A R Guadalupe[1,3]
[1] Department of Chemistry, P.O. Box 23346, University of Puerto Rico, Río Piedras Campus, San Juan, Puerto Rico, 00931-3346
[2] Department of Biology, P.O. Box CUH 100 Rd 908, University of Puerto Rico, Humacao Campus, Humacao, Puerto Rico, 00791

Email: anlupe55@gmail.com

Abstract. This report describes the synthesis of $[Ru(Fe\text{-}Phen)_2dppz](PF_6)_2$ (Ru-Fe complex) for a label-free approach to detect DNA hybridization. The Ru-Fe complex showed oxidation signals at +608 mV and +1192 mV corresponding to the RuII/III and FeII/III centers, respectively. We used the Ru-Fe complex and the Ferrocene covalently attached to the target to monitor the hybridization event of a 70-mer oligo immobilized in 10.3KD NHS-PS-NHS. The lowest target detectable concentration for the DNA fragment was around 0.4 μM.

1. Introduction

Waterborne and foodborne diseases are one of the principal public health problems worldwide. Our particular interest is on the development of nucleic acid biosensors (NAB) for the detection of pathogenic microorganisms in food and water samples. In a NAB, a ss-nucleic acid sequence is used as a probe to identify a complementary target sequence (the analyte) [1]. The detection of the hybridization process relies on the measurement of a redox-active molecule that interacts with the double strand. Two approaches to the use of redox labels can be mentioned. In one of these approaches, the redox label is conjugated to the probe or the target and the hybridization is signaled by measuring a current (or potential) or a resultant change for the case when the label is at the probe [1-4]. A second approach is the so-called label-free which is actually an indirect measurement of the hybridization process.

Ruthenium (II) polypyridyl complexes have received considerable attention because of their binding interaction with dsDNA [5-14]. $[Ru(bpy)2dppz]^{2+}$ and $[Ru(phen)2dppz]^{2+}$ (dppz = dipyrido[3,2-a: 2', 3'-c]-phenazine) have been reported as "DNA light-switching" molecules where their negligible luminescence in water is enhanced on the addition of dsDNA [6,11]. In these two complexes, the dppz ligand possesses an extended planar π-aromatic structure, which favors the complex intercalation between the base pairs of double helical DNA [5-14]. The interaction of $[Ru(L_2)ddpz]^{2+}$ complexes with dsDNA has been well studied [8,11-14] in solution. In this research, we report on the development of a NAB prototype using a polymer modified electrode together with a 70-mer probe selective to *Salmonella* as probes and Ruthenium-Ferrocene (Ru-Fc) bi-metallic

[3] To whom any correspondence should be addressed.

complex as a label (scheme 1). The results from these experiments were compared with prior experiments [15] where we used only ferrocene conjugated to the target to signal the hybridization event.

2. Methodology

2.1. Synthesis and characterization of the Ru-Fe complex (Scheme 1)

The dppz and Fc-Phen ligands were prepared as described by Dupureur and Barton [6,9] and by Zapata and coworkers [16], respectively. The cis-Ru(Fe-Phen)$_2$Cl$_2$ was synthesized following the procedure by Z. Ji and coworkers [17]. 17.41 mg (0.067 mmol) of $RuCl_3.xH_2O$ (x assumed to be 3) was refluxed with 70.56 mg (0.17 mmol) of Fe-Phen and 21.85 mg (0.52 mmol) of LiCl for 8 h in 10 mL of DMF under continuous stirring. The crude product was purified by suspending it in 200 mL of water:ethanol (1:1) and refluxing the mixture until all solids had dissolved (~30 min) to form a deep brown red solution. This solution was filtered and rotoevaporated to remove the ethanol in the presence of 1 g LiCl. 50 mL of acetone was added to precipitate the product. The crystals were filtered and rinsed with diethyl ether. The yield was 90%.

The resultant complex Ru(Fe-Phen)$_2$Cl$_2$ (99.61 mg, 0.11 mmol) and dppz (28.54 mg, 0.10 mmol) were refluxed overnight in 30 mL of DMF under N_2(g) and with continuous stirring. The volume was reduced, the solution cooled, and aqueous NH_4PF_6 added. The resulting red precipitate was filtered and rinsed with acetone followed by diethyl ether. The product was purified by size exclusion chromatography with Sephadex LH-20 and methanol:acetonitrile (1:1) as the eluent. The yield was 66%. MS-TOF, ^{1}H-NMR, UV-Vis and OSWV were used to characterize the ligands and the complexes. The electrochemistry was done in CH_3CN/ 0.1 M TBAP.

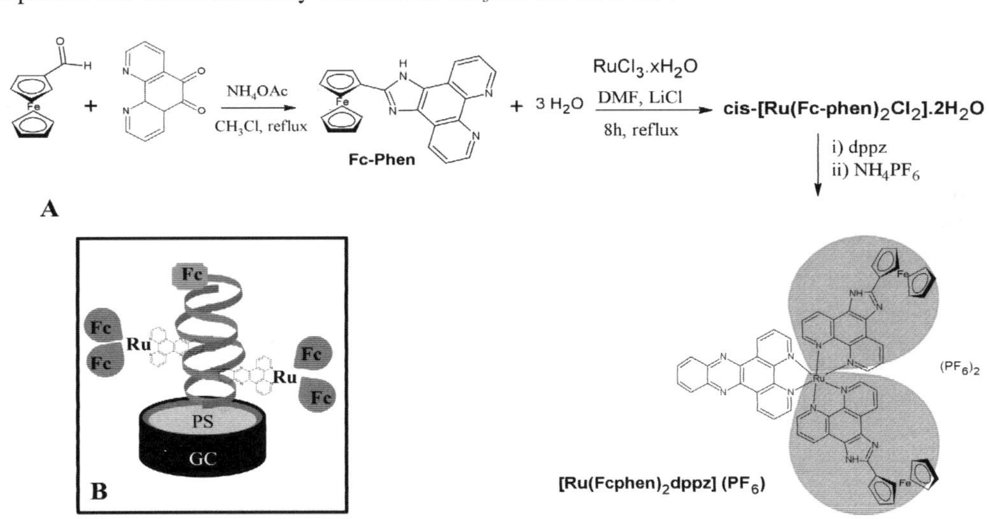

Scheme 1. Ru-Fc bimetallic intercalator and sandwich design of the DNA hybridization detection.

2.2. Preparation and calibration of a 70-mer DNA biosensor

Glassy carbon electrodes were cleaned and modified with a 10.3 KD NHS-PS-NHS polymer film as previously published [15]. A 20 μL of 7 μmol/L of 70-mer oligo (probe) solution was deposited at the polymer film and incubated at RT for 12 h. The probe-modified film was rinsed with 5 mL of deionized water for 5 min. The concentration and purity of the oligo solutions was previously determined by measuring the absorbance at 260 and 280 nm. A sample of pure ss-DNA with an absorbance equals to one at 260 nm contains approximately 33 mg/mL of DNA [15].

The hybridization of the immobilized probe was studied by looking at the redox process of (Scheme 1): (1) ferrocene covalently attached to the complementary 70-mer target sequence [15], and

(2) the Ru-Fe complex intercalated within the double strand. A 20 μL of the target solution was deposited at the probe-modified film, incubated at 70°C for 5 min and left at RT for 12 h. The film was rinsed successively with 5 mL of PBS (pH 6.9). An OSWV was run in PBS/0.103M NaCl (μ=0.271 M) from 200 to 1300 mV. The sensor was exposed to a 1.58 x 10^{-4} M Ru-Fe solution in PBS/0.103 M NaCl for 4 hr. This solution was prepared by mixing a 50 μL aliquot of a 7.9 x10^{-4} M Ru-Fe acetonitrile solution with 200 μL of PBS/0.103 M NaCl. OSWV followed under the same experimental conditions. Calibration curves were constructed varying the target concentration from 0.8 μM to 0.4 μM. The current density from the oxidation of the ferrocene in the ferrocene-labeled target and the ruthenium in the bimetallic complex were used for the construction of the calibration curves.

3. Results and Discussion

3.1. Characterization of the compounds. The electrochemical behavior of 1.3 mM Fe-Phen and 0.68 mM Ru-Fe complex was studied in CH_3CN/0.101 M TBAP with OSWV. The Ru-Fe complex showed oxidation signals at +608 mV and +1192 mV corresponding to the $Ru^{II/III}$ and $Fe^{II/III}$ centers, respectively (see Figure 1A-B). The Fe-Phen ligand showed an oxidation potential at 548 mV.

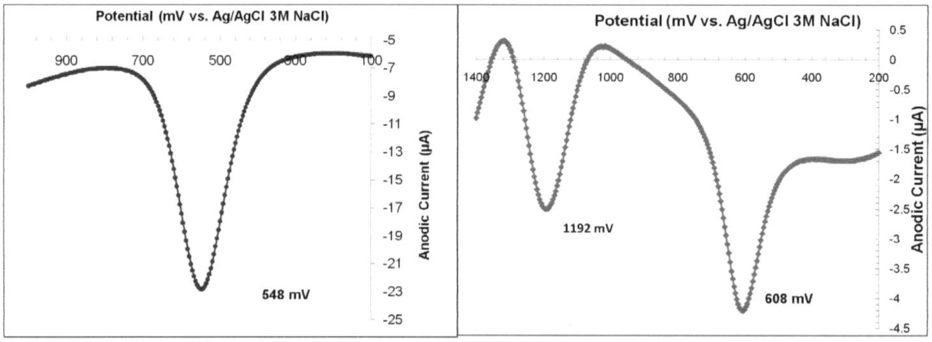

Figure 1. Square wave voltammetry of 1.3 mM Fc-Phen (blue) and 0.68 mM [Ru(Fc-phen)$_2$dppz](PF$_6$)$_2$ (red) between 100 and 1400 mV in acetonitrile 0.101 M TBAP. Glassy Carbon electrode, nichrome wire and Ag/AgCl (3 M NaCl) were used as working, counter and reference electrodes respectively.

3.2. Calibration of the 70-mer DNA biosensor.
Figure 2A (a-c) shows a sequence of voltammograms where the 70-mer target was hybridized with its complementary 70-mer probe immobilized at the 10.3KD NHS-PS-NHS modified glassy carbon electrode. The hybridization event was followed using the ferrocene-labeled target (2A-b) and the Ru-Fe complex (2A-c). As expected, we did not observe a redox signal for the polymer neither the probe (2A-a) because they are not electroactive in the potential interval studied. The voltammogram in 2A-b shows a signal at +336 mV resulting from the ferrocene oxidation [15]. In 2A-c, the voltammogram shows anodic peaks at +1092 mV and +532 mV after exposing the modified electrode surface to the Ru-Fe complex solution. The appearance of the two oxidation peaks confirms the interaction of the Ru-Fe complex with the immobilized dsDNA.

Calibration curves for the modified electrode using both detection schemes are shown in Figure 2B. The target concentration was varied from 0.8 μM to 0.4 μM. The current density of the Fe(II) oxidation in the ferrocene label and the Ru signal in the Ru-Fe complex were reported for each single point and used to construct the calibration plot. The shape of the curve has similar features, however the raising part of the 2B-b curve is steeper compared to curve 2B-a. The lowest target detectable concentration was around 0.4 μM for both cases. The standard deviation of each curve points varied between 0.25 μM to 0.01 μM for the ferrocene signal and 0.57 μM to 0.05 μM for Ru signal.

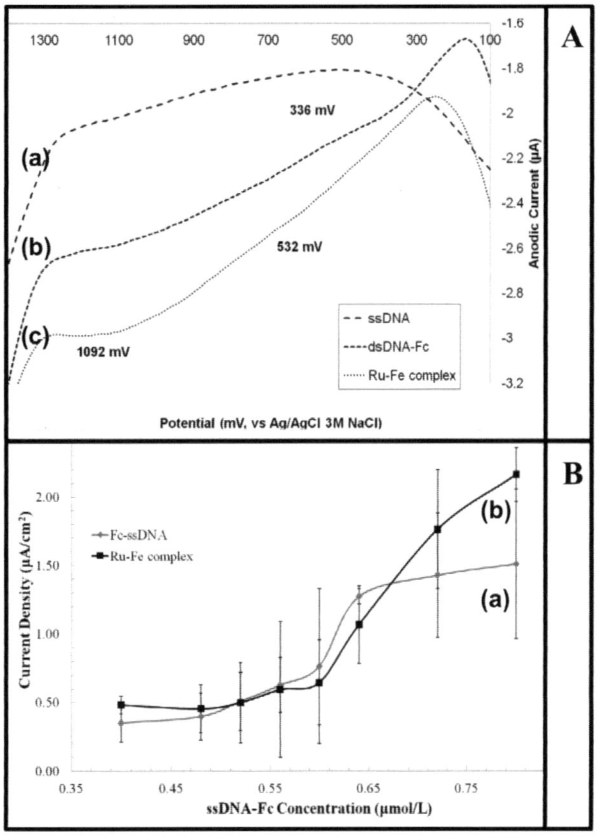

Figure 2. A: A 10.3 KD NHS-PS-NHS modified glassy carbon electrode with probe-target hybridization for the 70 mer. The hybridization event was followed using the ferrocene-labeled target (b) and the Ru-Fe complex (c). B: Calibration curves for a 10.3 KD NHS-PS-NHS modified glassy carbon electrode with a 70 mer oligo and its complementary target in 0.070 M PBS/ NaCl (μ = 0.271 M; pH 6.9) followed by the ferrocene-labeled target and the Ru-Fe complex. The target concentration was varied from 0.8 μM to 0.4 μM.

4. Conclusions

We synthesized and successfully characterized the ligands (Fe-phen and dppz) and the bimetallic Ru-Fe complex. The Fc-Phen showed the characteristic $Fe^{II/III}$ process of ferrocene. Ru-Fe complex showed the expected $Ru^{II/III}$ and $Fe^{II/III}$ redox processes. A 70-mer DNA biosensor using a polymer modified electrode anchoring and sensing platform and the Ru-Fe complex as the electrochemical label was constructed and calibrated. OSWV confirmed the Ru-Fe complex interaction with the immobilized dsDNA. The lowest target detectable concentration was above 0.4 μM. Our results show that it is possible to use metal complexes as DNA intercalators when appropriately designed for this purpose. Current work focuses on the sensor platform and the signal to noise ratio optimization by increasing the polymer hydrophilicity and thus the diffusion of molecules within the film. This sensor could be used to study the hybridization of complementary oligo sequences and their detection in real samples.

Acknowledgements
We acknowledge the kind support of NSF-PREM (DMR-0934195), NIH-MARC (5T34GM008156-20) and NIH-RISE (2R25GM061151-09). We also thank the technical assistance of the UPR, Río Piedras Materials Characterization Center.

5. References

[1] Wang J, Li JH, Baca AJ, Hu JB, Zhou FM, Yan W and Pang DW 2003 *Anal. Chem.* **75** 3941.
[2] Le Flocha F, Ho HA, Hurding LP, Bedard M, Neagu RP and Leclere M 2005 *Adv. Mater.* **17** 1251.
[3] Muñoz LS, Guadalupe AR and Vega EZ 2005 *J. Biotechnol.* **118** 233.
[4] Nakayama M, Ihara T, Nakano K and Maeda M 2002 *Talanta* **56** 857.
[5] Kerman K, Ozkan D, Kara P, Meric B, Gooding JJ and Ozsoz M 2002 *Anal. Chim. Acta* **462** 39.
[6] Friedman AR, Chambron JC, Sauvage JP, Turro NJ and Barton JK 1990 *J. Am. Chem. Soc.* **112** 4960.
[7] Harstborn RM and Barton JK 1992 *J. Am. Chem. Soc.* **114** 5919.
[8] Haq I, Lincoln P, Suh D, Norden B, Chowdhry BZ and Chaires JB 1995 *J. Am. Chem. Soc.* **117** 4788.
[9] Dupureur CM and Barton JB 1997 *Inorg. Chem.* **36** 33.
[10] Chem W, Turro C, Friedman LA, Barton JK and Purro NJ 1997 *J. Phys. Chem. B.***101** 6695.
[11] Olson EJC, Hörmann HA, Jonkman AM, Arkin MR, Stemp EDA, Barton JK and Barbara PF 1997 *J. Am. Chem. Soc.* **119** 11458.
[12] Greguric A, Graguric ID, Hambley TW, Aldrich-Wrigth JR and Collins G 2002 *J. Chem. Soc. Dalton Trans.* **6** 849.
[13] Biver T, Cavazza C, Secco F and Venturini M 2007 *J. Inorg. Biochem.* **101** 461.
[14] Hu L, Bian Z, Li H, Han S, Yuan Y, Gao L and Xu G 2009 *Anal. Chem.* **81** 9807.
[15] Díaz M, Rosado A, del Pilar J, Arias M and Guadalupe AR 2011 *Electroanalysis.* **23** 1830.
[16] Zapata F, Caballero A, Espinosa A, Tarraga A and Molina P 2008 *J. Org. Chem.* **73** 4034.
[17] Ji Z Huang SD and Guadalupe AR, 2000 *Inorg. Chim. Acta.* **305** 127.

FIA-automated system used to electrochemically measure nitrite and its interfering chemicals through a 1-2 DAB / Au electrode: gain of sensitivity at upper potentials

F L Almeida[1,3], **S G dos Santos Filho**[1] **and M B A Fontes**[2]
[1]Laboratório de Sistema Integráveis, Universidade de São Paulo, São Paulo, Brazil.
[2]Faculdade de Tecnologia de São Paulo – FATEC-SP/CEETEPS, São Paulo, Brazil.

Email: falmeida@lsi.usp.br

Abstract. The measurement of nitrite and its interfering-chemicals (paracetamol, ascorbic acid and uric acid) was performed employing a Flow-injection Analysis (FIA) system, which was automated using solenoid valves and air-pump. It is very important to quantify nitrite from river water, food and biologic fluids due to its antibacterial capacity in moderated concentrations, or its toxicity for human health even at low concentrations (> 20 μmol L^{-1} in blood fluids). Electrodes of the electrochemical planar sensor were defined by silk-screen technology. The measuring electrode was made from gold paste covered with 1-2 cis Diaminobenzene (DAB), which allowed good selectivity, linearity, repeatability, stability and optimized gain of sensitivity at 0.5 $V_{Ag/AgCl\ Nafion^{®}\ 117}$ (6.93 μA mol^{-1} L mm^{-2}) compared to 0.3 $V_{Ag/AgCl\ Nafion^{®}\ 117}$. The reference electrode was obtained from silver/palladium paste modified with chloride and covered with Nafion® 117. The auxiliary electrode was made from platinum paste. It was noteworthy that nitrite response adds to the response of the studied interfering-chemicals and it is predominant for concentrations lower than 175 μmol L^{-1}.

Keywords: nitrite measurement, 1-2 DAB, amperometric sensor and electrochemical potential

1. Introduction

From the 70's and 80's until now, the micro-fabrication, silkscreen and microelectronic technologies have been employed to fabricate nitrite biosensor and electrochemical sensors [1-3]. In recent years, the electrochemical techniques have been largely employed in sensors due to the easy setup, versatility in modification of the surface electrodes, and, especially, because they allow direct amperometric measurements with lower detection limits [2,4].

The need for quantitative measurement of nitrite (NO_2^-) has increased due to its toxicity potential for human health (if > 21.7 μmol l^{-1} in potable water) [5] and environmental security (< 21.7 μmol l^{-1} in river water) [6]. Nitrites can react with amines in stomach forming nitrosamines, which are known to be carcinogenic [4,7,8].

Nitrite concentrations in blood are in the range from undetectable to 20 μmol l^{-1} [9]. If nitrite is administrated at low levels (< 1 μmol l^{-1}), it can significantly reduce cardiac infarct size by approximately 50-67% [10]. On the other hand, if nitrite is present at high levels, it can react with hemoglobin, forming methemoglobin (reaction 1), which has no oxygen carrying ability [11,12].

[3] To whom any correspondence should be addressed.

$$4NO_2^- + 4HbO_2 \rightarrow 4NO_3^- + 2Hb_2O + O_2 \qquad (1)$$

Nitrite, as key species in the global nitrogen cycle [13,14], is present in soils, drinking and river waters, foods and physiological systems. Also, it is employed as food preservative against poisoning by microorganisms such as *clostridium botulinum* and it appears in water during chlorination.

Nitrite's anion is of interest as one Nitric Oxide (NO) precursor in biological systems, and its indirect detection has recently been investigated since it is an important element associated to some physiological events, like vasodilatation, neurotransmission [15], inflammation and cell proliferation. Reaction 2, as follows, illustrates the formation of NO, which may occur in the human body [4].

$$2NO_2^- + 2O_2 + 6H^+ \rightarrow 2NO + H_2O_2 + 2H_2O + O_2 \qquad (2)$$

The 1-2 cis Diaminobenzene (1-2 DAB) film was chosen as suggested by previous studies [2,16,17] that had shown an increase of selectivity to the nitrite and a decrease of the interfering-chemicals signals for uric acid (UA), ascorbic acid (AA) and paracetamol (PA) at maximum concentrations used as therapeutic doses in blood of a patient [17] 0.52 mmol l^{-1}, 0.11 mmol l^{-1} and 0.15 mmol l^{-1}, respectively.

In addition, we have studied the sensor integrated to an Automated Flow-injection Analysis (FIA-automated) System in order to improve repeatability and stability [18]. In Intensive Therapy Unities, it is important to control the nitrite concentration to avoid patient intoxication due to diet, accelerated metabolism or medicaments.

2. Procedures and Methods

2.1. Fabrication of the Sensor

The proposed sensor, composed by three electrodes, was silk-screened onto an alumina substrate of (17 x 43) mm. The measuring electrode was made with DuPont® 5142 gold paste, the reference electrode, with Pd/Ag DuPont® 6146 paste modified for the chlorination and covered with a Nafion® 117 membrane and the auxiliary electrode, with ESL 5545-LS platinum paste. The geometric areas were 4.5 ± 0.2 mm^2, 4.2 ± 0.7 mm^2 and 59.8 ± 0.7 mm^2; for measuring, reference and auxiliary electrodes, respectively. These can be observed in figure 1. In addition, Au paste and DuPont® 9615 were utilized to define connection lines / contacts and electric isolation, respectively.

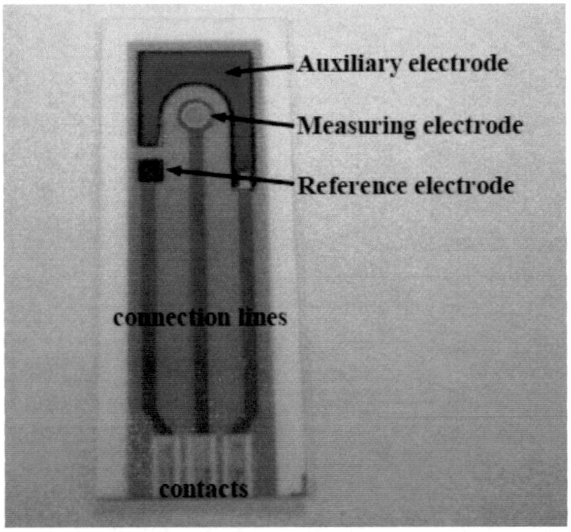

Figure 1. Schematic of the planar sensor with the respective electrodes.

2.2. Cleaning of the Electrodes

After the fabrication, it was necessary to activate the electrodes surface to guarantee the increase of sensitivity. In other words, a standard cleaning procedure (physic, chemical and electrochemical) was performed as reported in [17] to clean the gold surface. At first, the surface was polished with 0.05 μm gel alumina (thirty 8-shape movements), then, it was rinsed in acetone and alcohol, and, finally, Cyclic Voltametry (CV) was performed by applying 90 voltage cycles from -0.9 to +0.9 $V_{Ag/AgCl\ 3\ M\ KCl}$ (scanning rate of 0.3 V s^{-1}) in a 0.1 mol l^{-1} H_2SO_4 solution. The results of cleaning allowed maximum repeatability and reproducibility of the measuring electrode.

It was noteworthy that black spots on the electrode surface were observed under optical microscopy before been cleaned and that they were removed after cleaning, as shown in figure 2.

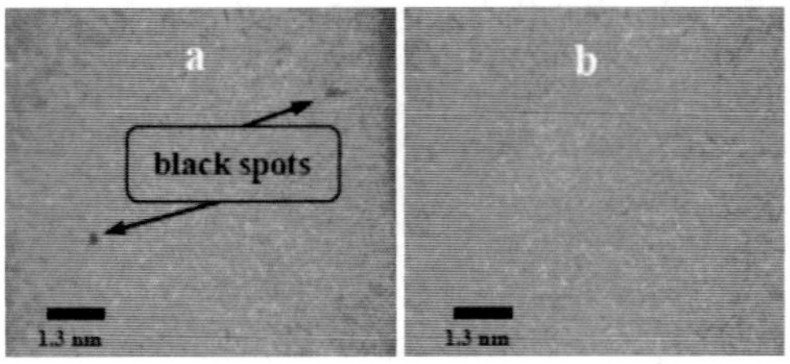

Figure 2. Optical microscopy images of the gold electrode surface, a) before cleaning procedures and b) after chemical cleaning procedures.

Normally, during the fabrication process of the electrochemical sensors or biosensors, it is convenient to deposit a conductor film above the working electrode surface to decrease the influence of the interfering elements and to increase the measured signal. In our work, 1-2 cis Diaminobenzene (DAB) was chosen because it fulfils this requirement and due to its previous use [16,18-20] with selectivity for nitrite. In addition, we have studied this membrane with Automated Flow-Injection Analysis (FIA) Technique in order to improve repeatability, reproductively, stability and linearity [18].

2.3. Characterization

CV characterization of the electrodes allows us to establish functionality based on the current measurement. The CV technique using a PalmSens voltammetric instrument was employed to obtain the effective area in solution for the bare-gold measuring electrode, and to evaluate its stability.

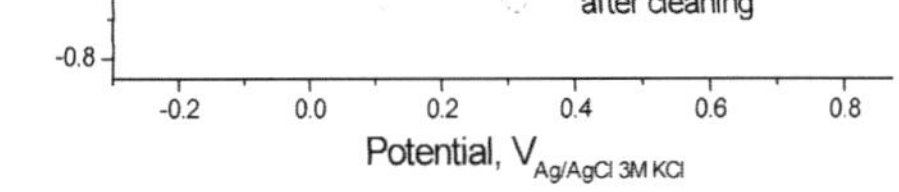

Figure 3. Cyclic voltammograms to extract peak current before and after cleaning, and, so, to calculate the effective area of the measuring electrodes.

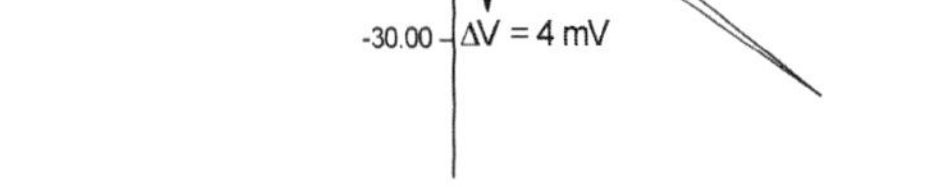

Figure 4. Reversibility curve of the Ag/AgCl Nafion® 117 film as the reference electrode.

A solution of 0.1 mol l^{-1} potassium ferricyanide in 0.5 mol l^{-1} KCl was used with the following parameters: 6 cycles of a voltage ramp varying from +0.8 to -0.2 $V_{Ag/AgCl\ 3M\ KCl}$ with scanning rate of 0.1 V s^{-1} (figure 3).

The effective area of 4.33 mm^2 for the measuring electrode was obtained from Randles-Sevcik's equation and with the aid of the CV in figure 3 [18,21]. One observes a decrease of 3.7 % in relation geometric area due to electric isolation layer around gold that reduces area.

To obtain a AgCl film on the reference electrode, a chlorination reaction was performed in a 0.1 mol l^{-1} HCl solution at a fixed potential of 0.86 VAg/AgCl 3M KCl (0.5 mA cm^2) during ~34 min (integrated charge of 950 mC for 5 μm of AgCl) followed by a confinement in serum physiologic (NaCl 0.9% and pH = 7.3) during three days.

Following, a 3 mol l^{-1} NaCl solution was employed to perform the tests of reversibility of the reference electrode. First, the Ag/AgCl junction was stabilized by using a voltage ramp from -0.3 to +0.3 $V_{Ag/AgCl\ 3M\ KCl}$ (50 cycles with scanning rate of 0.3 V s^{-1}); then, the reversibility test was performed for a voltage ramp from -0.5 to +0.5 $V_{Ag/AgCl\ 3M\ KCl}$ (6 cycles with scanning rate of 0.1 V s^{-1}), as shown in figure 4.

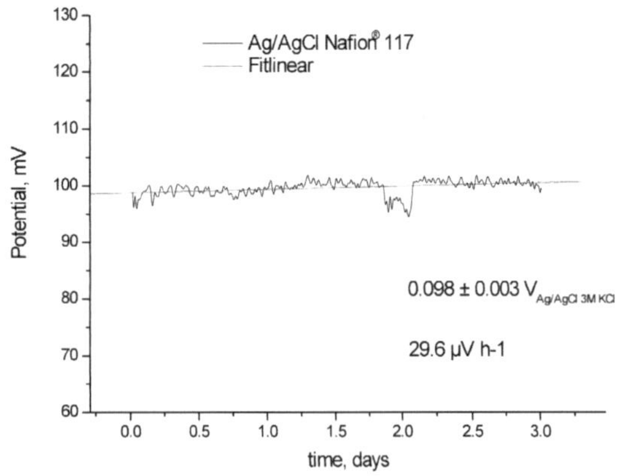

Figure 5. Potential evolution test as a function of the time of Nafion® 117.

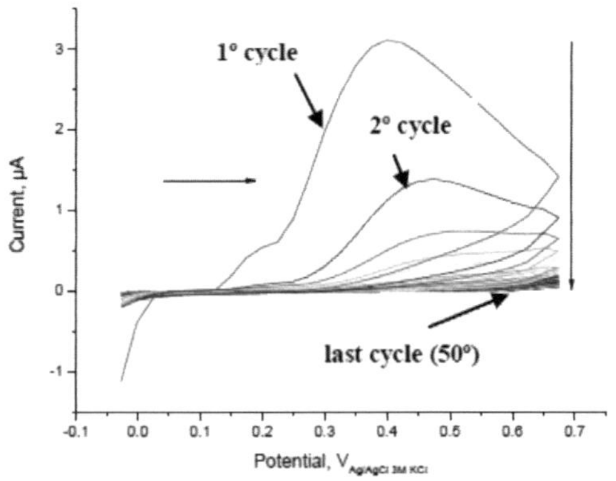

Figure 6. Electrodeposition of the 1-2 DAB film by CV (50 cycles from 0 to +0.65 $V_{Ag/AgCl\ 3M\ KCl}$ with scanning rate of 0.1 V s^{-1}).

Figure 5 shows another test. In this case, the voltage behavior of the Ag/AgCl Nafion® 117 film. The resulting average electrical potential was 98 ± 3 $mV_{Ag/AgCl\ 3M\ KCl}$ with a degradation of 29.6 μV h^{-1}.

2.4. Surface Modification of the gold electrode

5 mmol l^{-1} 1-2 DAB was added to 0.05 mol l^{-1} a phosphate buffer (pH = 7.2), which was purged during 30 min in nitrogen gas. Following, the 1-2 DAB film was electrodeposited over the gold surface using CV in PalmSens (50 cycles from 0 to +0.65$V_{Ag/AgCl\ 3M\ KCl}$ with scanning rate of 0.1 V s^{-1}) as shown in figure 6.

1-2 DAB film electro-activated onto gold surface contributes to decreasing the reactivity and increasing the stability and repeatability of the measured FIA signals.

2.5. FIA-automated System

The FIA system is composed of solenoid valves, discard, analysis-cell, air-pump, electrical and fluidic connections, PalmSens potentiostat and analytes. The experimental set-up is shown in figures 7 and 8. It is noteworthy the analysis-cell (at the lower right corner) and control computer (at the upper left corner) in figure 8.

The physiologic serum (NaCl 0.9 % and pH = 7.3) was employed as solvent or as cleaning and carrier solution during the measurements. The concentrations performed for nitrite were 50, 100 and 250 μmol l^{-1}, and the interfering-chemicals were UA (0.55 mmol l^{-1} diluted in 0.1 mol l^{-1} NaOH), AA (0.11 mmol l^{-1}) and PA (0.17 mmol l^{-1}) with concentration chosen at the maximum therapeutic doses in blood of a patient [17].

Figure 7 shows the operational scheme of the FIA-automated System, developed for measurement of nitrite and their interfering-chemicals.

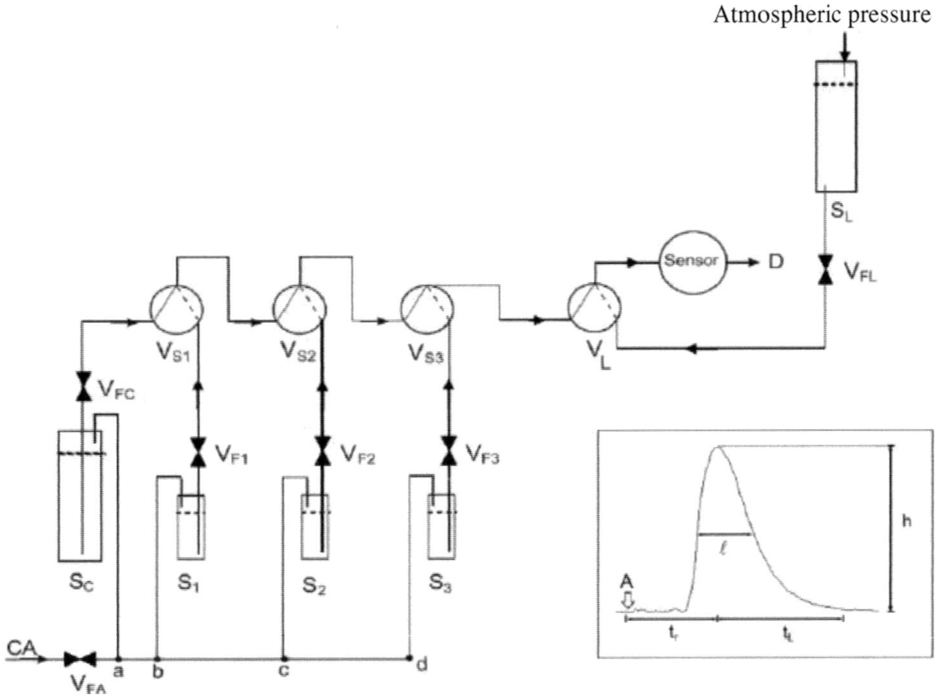

Figure 7. FIA-automated composed by the solenoid valves (V_{S1}, V_{S2}, V_{S3} e V_L) and flow control valves (V_{FC}, V_{F1}, V_{F2} e V_{F3}) for injection of analysis solutions (S_1, S_2 e S_3), cleaning solution (S_L), carrier solution (S_C), sensor, discard (D) and FIA signal.

It is identified in the scheme of the FIA-automated System:

a, b, c, d: dots of air confluence.
S_C: Carrier Solution.
S_L: Cleaning Solution.
S_1: Analyte Solution 1.
S_2: Analyte Solution 2.
S_3: Analyte Solution 3.
V_{S1}, V_{S2}, V_{S3} e V_L off, it flows S_C.
V_{S1}: Solenoid Valve 1 on, it injects S_1.
V_{S2}: Solenoid Valve 2 on, it injects S_2.
V_{S3}: Solenoid Valve 3 on, it injects S_3.
V_L: Solenoid Valve of cleaning on, it flows S_L by action of gravitational force.
V_{FC}: Valve to set flow S_C.
V_{FL}: Valve to set flow S_L.
V_{F1}: Valve to set flow S_1.
V_{F2}: Valve to set flow S_2.
V_{F3}: Valve to set flow S_3.
V_{FA}: Valve to set air flow, it allows to drain air excess.
CA: air compressor employed for pumping by pressure, carrying of solutions.

Figure 8 is a picture of the FIA-automated System indicating the solenoids and flow valves to control the nitrite, interfering-chemicals solutions, the cleaning and the carrier solutions during the FIA process.

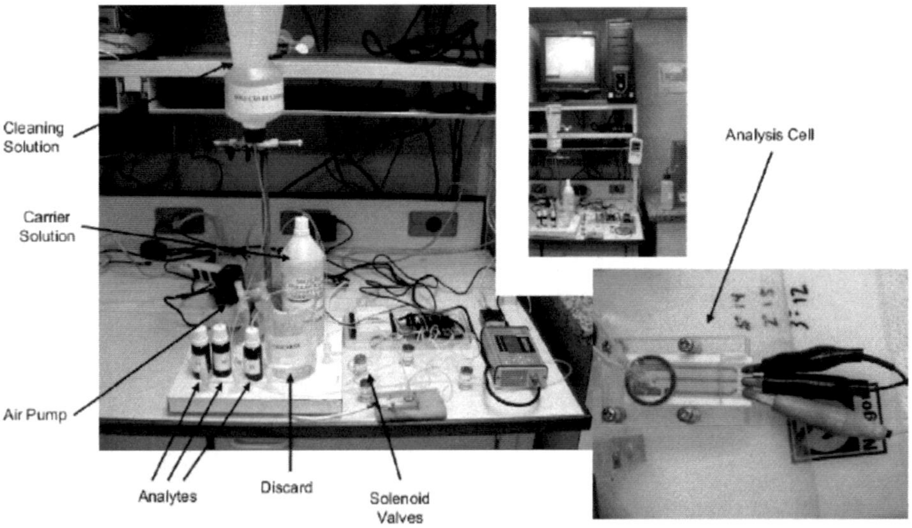

Figure 8. Automated FIA system to measure nitrite and interfering-chemicals.

The picture of the experimental setup in figure 8 also shows analysis cell, air pump, analytes, analysis-cell, fluidic and electrical connections and discard. This equipment was designed to simulate blood flow utilizing physiologic serum (NaCl 0.9 % and pH = 7.3) as carrier and cleaning solution in the FIA-automated system.

2.6. Determination of the electrochemical potentials for FIA

Differential Pulse Voltammetry (DPV) technique [22] was employed to obtain the electrochemical potentials to be used in FIA in order to obtain the concentrations of nitrite, AA, UA and PA at constant potentials. Figure 9 illustrates the start and the peak potential for nitrite reduction.

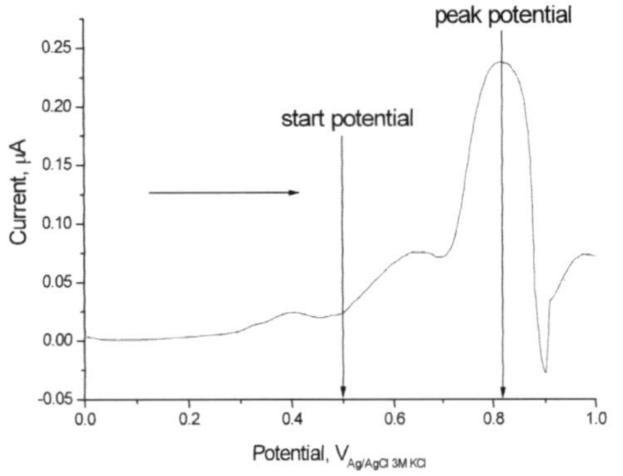

Figure 9. Standard DPV measurement showing the start and the peak current and potential for nitride reduction.

Table 1 shows the peak current, the start and the peak potentials of nitrite, AA, UA and PA for concentrations of 2.00 mmol l^{-1}, 0.44 mmol l^{-1}, 2.00 mmol $l^{-1}$and 0.68 mmol l^{-1}, respectively. The concentrations of AA, UA and PA were chosen as the maximum therapeutic doses in blood of a patient, here defined as the worst condition to be achieved in order to allow the measure of the nitrite concentration without the interference of AA, UA and PA.

Based on the peak currents presented in Table 1 and aiming to perform FIA process at constant potentials, 0.3 and $0.5V_{Ag/AgCl\ Nafion®117}$ were arbitrarily chosen to allow neglecting the measured current for AA, UA and PA compared to the corresponding for the nitrite concentration.

Table 1. Electrochemical potentials obtained for nitrite and its interfering-chemicals.

Metabolites	Concentration Mmol l^{-1}	Peak Potential $V_{Ag/AgCl\ Nafion®\ 117}$	Start Potential of $V_{Ag/AgCl\ Nafion®\ 117}$
Nitrite	2.00	0.880	0.500
Ascorbic Acid (AA)	0.44	0.292	0.172
Uric Acid (UA)	2.00	0.472	0.352
Paracetamol (PA)	0.68	0.602	0.372

3. Results and Discussion

Using the procedure as described in the experiment, FIA measurements were performed. Graphs "a" and "b" in figure 10 correspond to the FIA response of the interfering-chemicals (AA, PA and UA). It was noteworthy that AA and paracetamol presented similar pulsed-current signals at $0.3V_{Ag/AgCl\ Nafion®117}$ above the nitrite signal in figure 10(c). On the other hand, for $0.5V_{Ag/AgCl\ Nafion®117}$, AA, PA and UA signals were lower than the nitrite signal compare figure 10(b) with figure 10(d)). It means that the interfering-chemicals can be neglected compared to nitrite.

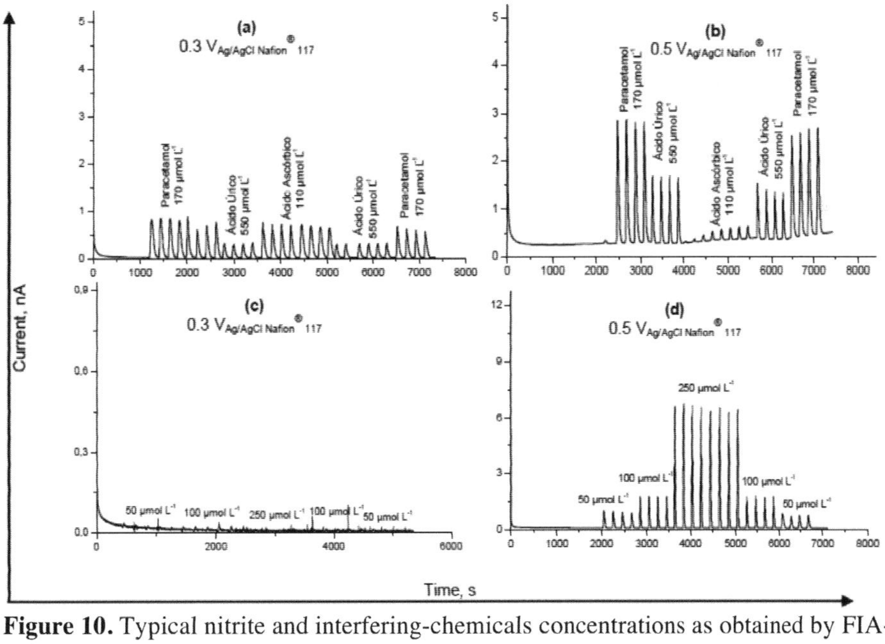

Figure 10. Typical nitrite and interfering-chemicals concentrations as obtained by FIA-automated at 0.3 and $0.5\ V_{Ag/AgCl\ Nafion®117}$.

From FIA, the curves of sensitivity shown in figure 11 could be plotted and the slope could be extracted and normalized concerning to the effective area of $4.33\ mm^2$. In addition, current signals of the interfering-chemicals compared to the nitrite were also evaluated from FIA (figure 12).

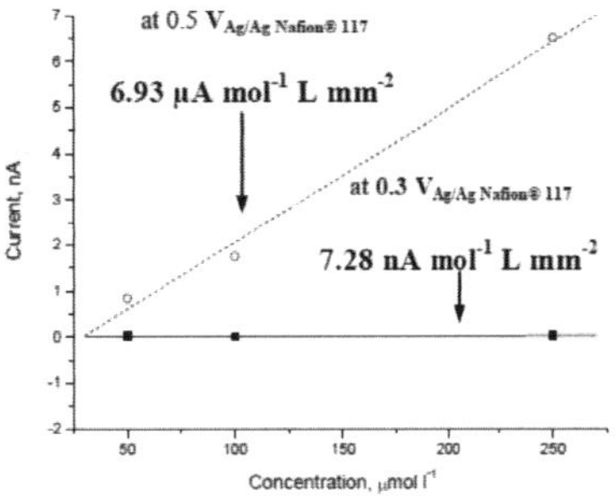

Figure 11. Curves of sensitivity for nitrite obtained at 0.3 and $0.5V_{Ag/AgCl\cdot Nafion®117}$, respectively.

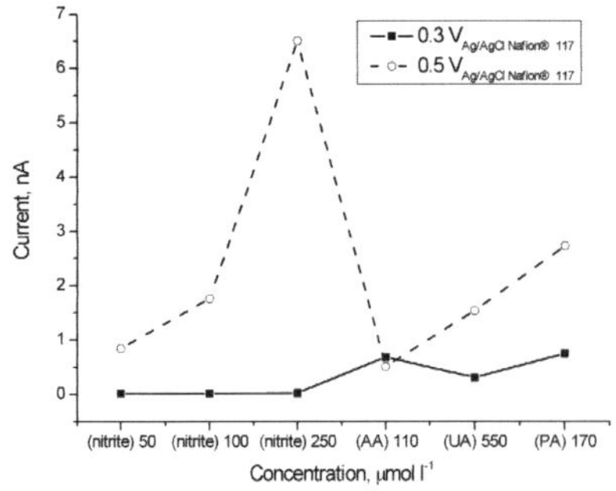

Figure 12. Current response for nitrite and its interfering-chemicals at 0.3 and $0.5V_{Ag/AgCl\ Nafion®117}$ potential.

The result above indicates that the nitrite signal is higher than the interfering-chemicals for 0.5 $V_{Ag/AgCl\ Nafion®\ 117}$. On the other hand, for 0.3 $V_{Ag/AgCl\ Nafion®\ 117}$, the nitrite signal is lower than the signals of the interfering-chemicals (AA, UA and PA). As a result, it is concluded that it is feasible to obtain the concentration or nitrite by FIA at 0.5 $V_{Ag/AgCl\ Nafion®117}$ in a selective way without substantial influence of UA, PA and AA.

4. Conclusions

The presented results indicated linearity, repeatability, stability and selectivity for nitrite measurement. It was demonstrated the feasibility of obtaining the concentration of nitrite by FIA at 0.5 $V_{Ag/AgCl\ Nafion®\ 117}$, in a selective way without substantial influence of UA, PA and AA. The obtained sensitivity for nitrite was on order of 6.93 μA mol^{-1} L mm^{-2} and 7.28 nA mol^{-1} L mm^{-2} for FIA performed at 0.5 $V_{Ag/AgCl\ Nafion®\ 117}$, and 0.3 $V_{Ag/AgCl\ Nafion®\ 117}$, respectively.

5. References

[1] Suzuki H 2001 *J. Electrochem. Soc.***12** 468
[2] Adhikari B 2004 *Prog. Polym. Sci.* **29** 699
[3] Stetter J R, Penrose W R, and Yao S 2003 *J. Electrochemical Soc.* **150** (2) S11
[4] Moorcroft M J, Davis J and Compton R G 2001 *Talanta* **54** 785
[5] Portaria Ministério da Saúde 518/2004 Brasília 2005
[6] CONAMA RESOLUÇÃO No 357 (17-03-2005)
[7] Yang S. et. al. 2010 *Sens. Actuators B* **B145** 762
[8] Rajeshwar K and Ibanez J G 1997 *Environmental Electrochemistry* **29** 1449
[9] Dejam A et al. 2005 *Blood* **15** 106 734
[10] Sinha S S, Shiva S and Gladwin M T 2008 *Trends Cardiovasc Med* **18** 163
[11] Greenberg L A et al. 1943 *J. Biol. Chem.* **151** 665
[12] Tomoda A, Tsujit A and Yoneyama Y 1981 Journal Biochemical 193 169
[13] Teske A. et al. 1994 *J. Bacteriol.* **176** 6623
[14] Melo Filho A B and Biscontini T M B 2004 *Ciên. e Tecn. de Alimen.* **24** 390
[15] Mattos L 2001 *Química Nova* **24** 200
[16] Almeida F L and Fontes M B A 2007 *ECStransactions M. T. and D* **9** 545
[17] Laschi S 2006 Medical Engineering & Physics 28 934
[18] Curulli A et al. 2001 *J. Electroanal. Chem.* **509(1)** 66
[19] Seyama T. 1988 *Chemical Sensor Technology* 1 chap1
[20] Curulli A et al. 1997 *Biosens. Bioelectron.* **12** 1043
[21] Mabbott G A 1983 *J. Chem. Educ.* **60** 697
[22] Buffle J and Tercier-Waeber M L 2005 *Trends Anal. Chem.* **24** 172

Fabrication of phosphorus doped polysilicon thin-film strain gauges using a 50 microns silicon substrate thickness

A L Siarkowski[1], B S Rodrigues and N I Morimoto
LSI - Integrated Systems Laboratory. EPUSP - Polytechnic School of University of São Paulo. Av. Prof. Luciano Gualberto, Trav. 3, 158, São Paulo, SP, Brazil. CEP 05508-900

E-mail: acacio@lsi.usp.br

Abstract. Strain gauges fabrication using phosphorus doped polysilicon thin-film resistors was performed. Strain gauges transducers were designed to measure the strain in load cells, using a Wheatstone bridge circuit configuration. The strain and sensitivity measurements results for load cells application are described. A linear response and excellent repeatability were obtained.

1. Introduction

Typical strain gauges are made from a heat-treated metallic foil bonded to a dielectric layer [1, 2], as showed in Figure 1. However, phosphorus doped polysilicon layer is often used as a piezo-resistive material, mainly because the temperature coefficient of resistance (TCR) can be set as zero by suitable adjustment of the doping concentration [3, 4], due to this convenient feature, these materials are used as sensor for precision mass measurement. Polysilicon thin films strain gauges have high accuracy, repeatability, reliability and low cost. These characteristics make this material particularly attractive in load cells elements applications.

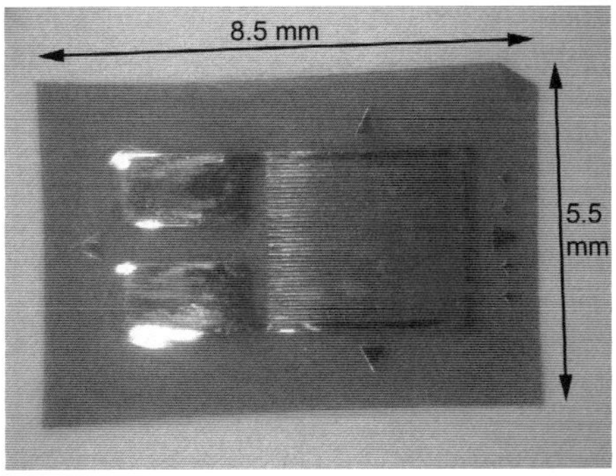

Figure 1. Conventional strain gauge made from a heat-treated metallic foil bonded onto a dielectric layer.

An important issue in this application is the bonding method for the strain gauge with the deformation surface of the sensor element (usually aluminum). Usually the strain gauges are positioned onto the areas, where the highest deformation is located on the load cells. The strain gauges

[1] To whom any correspondence should be addressed.

Published under licence by IOP Publishing Ltd

are connected to a Wheatstone bridge circuit configuration so when a load is applied in the load cell, the strain gauge detects the strain, and an unbalance in the Wheatstone bridge is observed. The resistance variation of the strain gauge is proportional to the mass of the load [3]. This paper shows the fabrication process for the phosphorus doped polysilicon thin film strain gauges and the optimization of phosphorous diffusion procedure as to enhance the sensor sensitivity. The fabricated devices were tested in load cells used in commercial mass measurement equipment.

1.1 Load Cells fundamentals

Nearly every electronic mass measurement equipment use load cells. These cells convert a force or mass into an electrical signal by transducing the strain into an electrical signal. As the surface of load cell becomes strained, the gauge stretches or compresses, changing its resistance proportionally to the applied load. Each strain gauge forms one leg of Wheatstone bridge, see Figure 2. When the excitation voltage E is applied to the circuit (AD), the output (BC) is a differential voltage proportional to the force on the load cell [6].

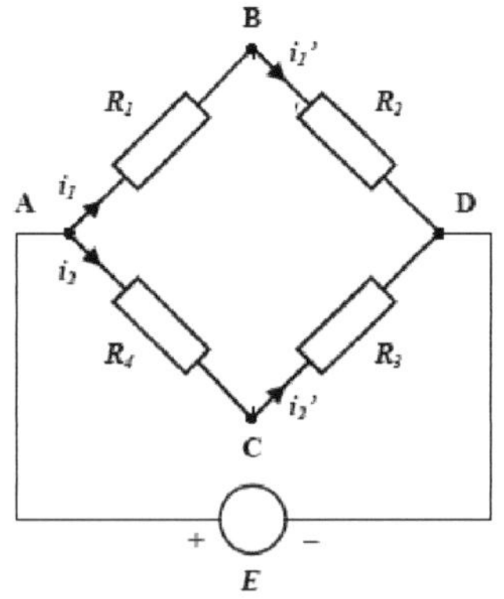

Figure 2. Electrical schematic of a Wheatstone bridge utilized in a load cell application [6].

Most load cells are specified in terms of maximum capacity and output rating. Maximum capacity defines the maximum load (in kg) to which the load cell can be stressed. After this value, the output voltage no longer changes. The sensitivity (output rating) is given in mV/V. The differential output voltage varies in a linear relation to the mass applied load cell. [6].

2. Fabrication and bonding of phosphorus doped polysilicon thin film strain gauges

Polysilicon thin films have sensitivity in between that of a metal foil and crystalline Si. However, the temperature coefficient of resistance, (TCR), which represents the extent of the effect of temperature changes, is significantly lower than that of crystalling Si strain gauges; it is typically less than 500 ppm/°C [7]. TCR is influenced by two factors: difference in the thermal expansions of the metal surface and the silicon strain gauge, and thermionic emission of carriers. In this study, the resistance change due to the thermionic emission was more of a factor in the gauge output determination than that by thermal expansion in the silicon strain gauges. The temperature effect can be compensated using a Wheatstone bridge circuit or by optimizing the fabrication process. Moreover, the design of the polysilicon strain gauges is straightforward due to the isotropic characteristics of these types of gauges as a result of the random grain distribution [1].

2.1 The fabrication process for the thin polysilicon strain gauge

A 100 oriented, p type, 10-20 Ohm.cm, 50μm thick silicon wafer was used as substrate. It was processed with the following steps:

- Chemical cleaning: The wafers were cleaned by standard RCA cleaning procedure to make their surfaces hydrophilic.

- Thermal oxidation: The wafers were thermally oxidized to grow a silicon oxide layer of 1μm thick.
- Polysilicon Deposition: A polysilicon thin film of 0.6μm thickness was deposited by Low Pressure Chemical Vapour Deposition (LPCVD) technique at 630°C.
- Phosphorus doping: A SOG (Spin on Glass) layer, with 1% or 3% phosphorus concentration, was deposited on the polysilicon film and the diffusion process temperature was carried out at 1150°C. The SOG layer was removed using a 1% HF solution.
- Patterns definition: The patterns of the designed strain gauges were defined by optical lithography and the polysilicon film was etched by a SF_6 plasma process. The photoresist was removed by O_2 plasma etching process.
- Metallization: An aluminum layer of 1μm thick was deposited by thermal evaporation. This aluminum layer was annealed at 420°C in inert gas ambient for 2 hours to improve the electrical contacts.
- Contacts definition: The contacts patterns were defined by optical lithography and etched in a H_3PO_4 solution. The photoresist was removed by acetone and 2-propanol solvents. Figure 3 shows an image of strain gauges with the electrical contacts patterns.
- Strain gauges dicer: The silicon wafer was diced by a disco saw to detach the individual strain gauges.

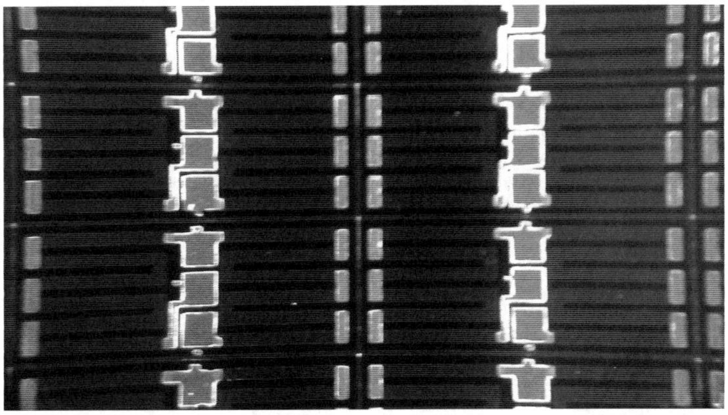

Figure 3. Image of the polysilicon strain gauges (1.2 x 3.0mm) with the aluminum electrical contacts.

2.2 Bonding on a metal cantilever beam and electrical wiring

Bonding between various types of metals and silicon devices can be implemented using organic adhesives or glass frit bonders. Although the strain gauges that are bonded using glass frit, have low hysteresis and high level of repeatability, one major disadvantage exists [2, 8, 9], it is difficult to bond a silicon strain gauge onto a metal cantilever beam using glass frit because the high-temperature hardening process during the glass-bonding process causes cracking due to the CTE (coefficient of thermal expansion) mismatch between the silicon gauge and metal cantilever beam. For the organic adhesives, the curing temperature is lower than glass frit bonders to prevent the film from cracking. To avoid difficulties, a high-performance epoxy resin was used to bond the polysilicon strain gauges onto a metal cantilever beam as showed in Figure 4.

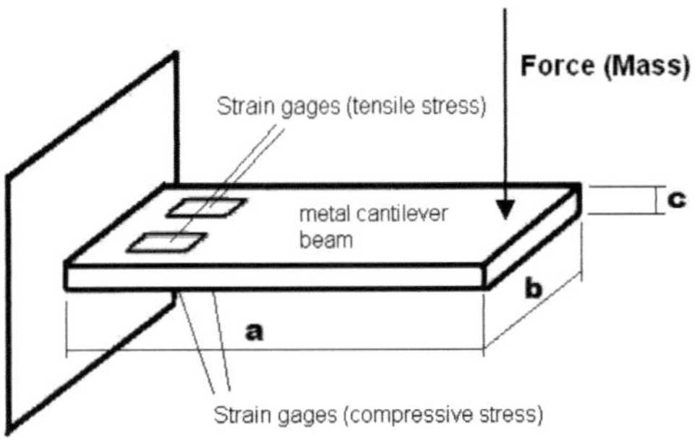

Figure 4. Metal cantilever beam used in the performance tests [1].

The metal and the cantilever beam dimensions (a x b x c) determines the maximum force (mass) that can be applied to the load as to avoid plastic behavior (not leading to hysteresis). The placement of four strain gauges in each load cell allows for the effective assembly of a Wheatstone bridge as shown in Figure 2. Notice that the strain gauges positioned at R_1 and R_3 have a positive resistance variation and R_2 and R_4 have a negative resistance variation [10].

The electrical contacts of the strain gauges can be made using a solder paste or wiring-bond processing. The Figure 5 shows an image of one strain gauge bonded using a high-performance epoxy resin and the electrical contacts were bonded using a silver conductive paste.

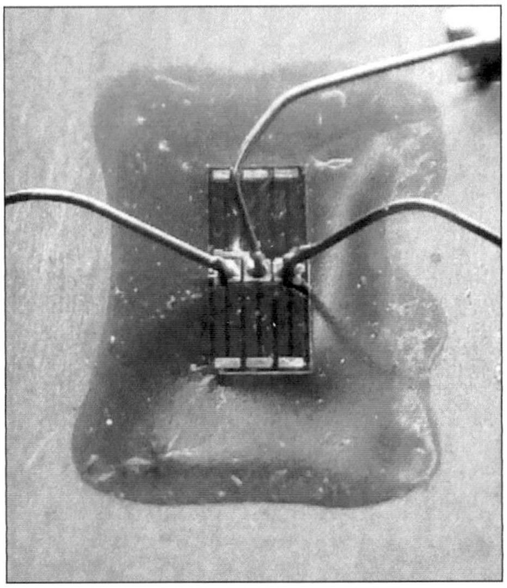

Figure 5. Strain gauge (1.2 x 3.0mm) bonded on aluminum cantilever beam using an epoxy resin and a silver conductive paste was used to perform the electrical contacts.

3. Experimental Results

3.1 Electrical characterization of the strain gauges.

3.1.1 *Measurements of sheet resistance:* The diffusion process changes the sheet resistance of the polysilicon film and it was characterized by a four-probe-meter. The main results of sheet resistance as a function of diffusion time and the concentration of the phosphorus dopant are shown in Table 1.

Table 1. Values of sheet resistance (Rsheet) of doped polysilicon film.

Phosphorus Concentration	Processing Time (minutes)	Rsheet (Ohms/square)
1%	20	150 ± 16
1%	30	42 ± 7
1%	40	14,4 ± 0,5
3%	70	10,7± 0,2

The variation of sheet resistance depends mainly on the thickness and doping uniformity of the polysilicon film, but is within acceptable values for the strain gauge designed.

3.1.2 Measurements of electrical resistance: The electrical resistance of the fabricated strain gauges was measured using an I-V analyzer, where a voltage of 1V was applied to strain gauges distributed in the silicon wafer. The average values for the fabricated devices are shown in Table 2.

Table 2. Values of electrical resistance of fabricated strain gauges.

Strain gauge identification	Resistance - Designed	Resistance - Measured
A	2 kΩ	2,10 ± 0,03 kΩ
B	1.2 kΩ	1,27 ± 0,02 kΩ
C	1.3 kΩ	1,34 ± 0,02 kΩ

The electrical resistance values were close to the designed resistance values, indicating that the polysilicon film deposition and phosphorus SOG difusion process results can be reproduced.

3.2 Electrical characterization of Wheatstone bridge in the load cell.
In the experimental arrangement, showed in Figure 4, an aluminum cantilever beam as a load cell was used. The arrangement was mounted on a stable and vibration free table. The input voltage was 5V and output voltage was monitored by a precision voltmeter as a function of the force (mass) applied on the structure. The results are shown in Figure 6.

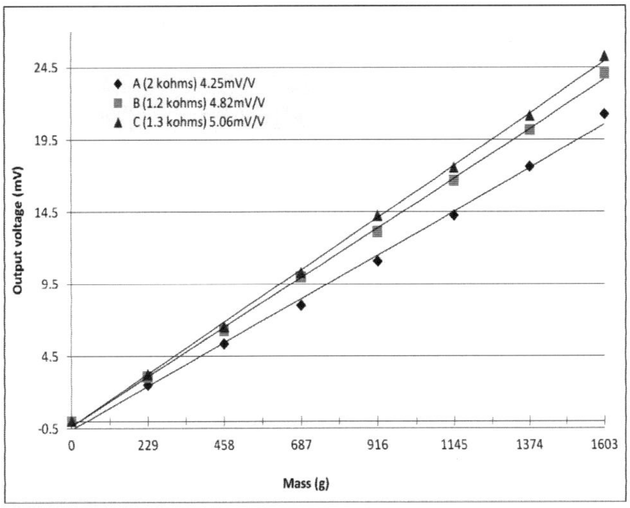

Figure 6. Wheatstone bridge output voltage in function of the force (mass) applied on the load cell.

A linear variation of Wheatstone bridge output voltage as a function of the force applied to the structure was observed with good correlation ($R^2 > 0.996$). So, the resistance of each strain gauge increased linearly with the tensile stress while it decreased with the compressive stress. In all cases, the sensitivity (output rating) for loads of 1.6 kg was higher than 4.25 mV/V (2.41 μV/V.g). The design of strain gauge C had longer lines in the same device and it was doped with a high phophorus concentration, resulting in a sensitivity of 5.06 mV/V (3.00 μV/V.g).

4. Conclusions

Strain gauges were designed and fabricated on 50 μm thick silicon wafers using conventional fabrication processes: thermal oxidation, chemical deposition of polycrystalline silicon (LPCVD), dopants diffusion, optical lithography, aluminum evaporation and wet and dry etching.

We demonstrated that the fabricated phosphorus doped polysilicon thin films strain gauges can be successfully used in a load cell sensor application.

The sheet resistivity of 10 ohms/square was adjusted in the phosphorus diffusion process and the strain gauges sensitivity depends on the length of the lines designed in the device.

The results of the electrical characterization of the Wheatstone bridges in the load cells show a linear variation of output voltage depending of the force (mass) applied to the structure with good correlations. The highest sensitivity (output rating) obtained was 5.06 mV/V (3.00 μV/V.g) for loads of 1.6 kg.

Acknowledgments

Authors would like to thanks NAMITEC, FAPESP, CAPES and CNPq for the financial support.

5. References

[1] Window A L 1992 *Strain Gauge Technology* Applied Science 2nd ed. (London: Elsevier).

[2] Khan A S and Wang X 2001 *Strain Measurements and Stress Analysis* Upper Saddle River (NJ: Prentice-Hall)

[3] Sivakumar K, Dasgupta N, Bhat K N and Natarajan K 2006 Sensitivity Enhancement of Polysilicon Piezo-resistive Pressure Sensors with Phosphorus Diffused Resistors *J. Phys.: Conf. Ser.* **34** 216-221

[4] Kim S C and Wise K D 1983 Temperature Sensitivity in Silicon Piezo-resistive Pressure Transducers *IEEE Trans. Electron. Devices* **30-7** 802-810

[5] Kim Y, Kim Y, Lee C and Kwon S 2010 Thin Polysilicon Gauge for Strain Measurement of Structural Elements *IEEE Sensors J* **10-8** 1320-1327

[6] Chiu T 2005 Incorporating the MSC1210 into Electronic Weight Scale Systems, Application Report of Texas Instruments

[7] Beeby S, Ensell G, Kraft M and White N 2004 *MEMS Mechanical Sensor* (Norwood, MA: Artech House)

[8] Leasure W A, Woodruff N and Gravel C 1971 Glass-bonding techniques for semiconductor strain gauge Experimental Mechanics *Proc. SESA Spring Meeting* **11-5** 235–240

[9] Knechtel R 2005 Glass frit bonding: A universal technology for wafer level encapsulation and packaging *Microsyst Technol* **12** 63–68

[10] Ciureanu P and Middelhoek S 1992 *Thin film resistive sensors* (Bristol: Institute of Physics)

Photodetector with integrated optical thin film filters

M A Vieira[1,3,4], M Vieira[1,2], P Louro[1,2], V Silva, A S Garção[2,3]
[1]Electronics Telecommunication and Computer Dept. ISEL, Lisboa, Portugal.
[2]CTS-UNINOVA, Quinta da Torre, Monte da Caparica, 2829-516, Caparica, Portugal.
[3]DEE-FCT-UNL, Quinta da Torre, Monte da Caparica, 2829-516, Caparica, Portugal.

E-mail: mv@isel.ipl.pt

Abstract. This paper reports on optical filters based on a-SiC:H tandem pi´n/pin heterostructures. The spectral sensitivity is analyzed. Steady state optical bias with different wavelengths are applied from each front and back sides and the photocurrent is measured. Results show that it is possible to control the sensitivity of the device and to tune a specific wavelength range by combining radiations with complementary light penetration depths. The transfer characteristics effects due to changes in the front and back optical bias wavelength are discussed. Depending on the wavelength of the external background and irradiation side, the device acts either as a short-or a long-pass band filter or as a band-stop filter. The output waveform presents a nonlinear amplitude-dependent response to the wavelengths of the input channels.

1. Introduction

Optical communication in the visible spectrum usually interfaces with an optoelectric device for further signal processing. There has been much research on semiconductor devices as elements for optical communication [4,5,6]. Multilayered Si/C structures based on amorphous silicon technology are expected to become reconfigurable to perform WDM optoelectronic logic functions. Filters are used to emphasize signals in a certain wavelength range and to reject signals in other ranges. They have a nonlinear magnitude-dependent response to each incident light wave. This nonlinearity provides the possibility for selectively removing and adding wavelength and can be used to boost signal after multiplexing or before demultiplexing.

An optoelectronic device converts light photons to electrons that mimic the light signal in such way that data transmitted by the light beam can be received and processed further with electrical circuits. These devices have one surface over which the optical signal shines. On our device, a multilayered Si/C heterostructure light [7, 8] can shine on the two surfaces namely back and front. This device acts as an optical filter when other fixed wavelengths superimpose the incident light data signal on the surface it shines with. By selecting a wavelength on either the red or blue part of the spectrum the device can be tuned as a filter and used as a wavelength division multiplexing-demultiplexing technique, WDM [9]. This device has been characterized with a model with its experimental verification. Here, we propose a photodetector with integrated optical thin film filters. By means of optical control signals applied to the device, the photonic function may be

[4] To whom any correspondence should be addressed.

modified giving reconfiguration. A new method is presented for optical routing using wavelength background processing techniques.

2. Device optimization and characterization

The sensor element is a multilayered heterostructure based on Si:H and a-SiC:H. The configuration shown in figure 1 includes two stacked p-i-n structures sandwiched between two transparent contacts. The thicknesses and optical gap of the front i'-(200nm; 2.1 eV) and back i-(1000nm; 1.8eV) layers are optimized for light absorption in the blue and red ranges, respectively. Spectral response measurements without and under different optical bias and frequencies were performed in order to test the devices sensitivity.

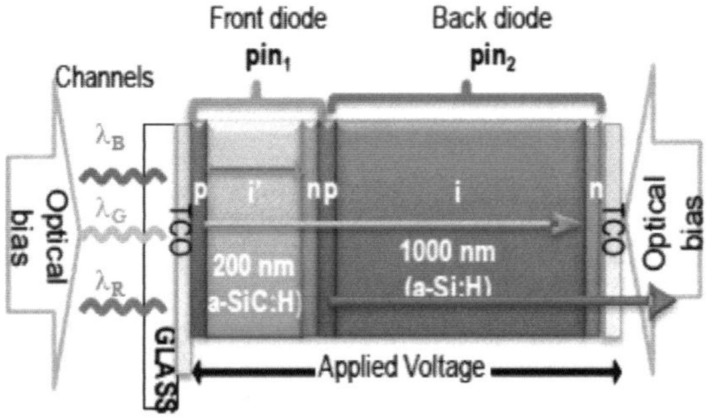

Figure 1. Device configuration and operation

The device operates within the visible range using as input color channels (data) the wave square modulated light (external regulation of frequency and intensity) supplied by a red (624 nm; 51 $\mu W/cm^2$), a green (526 nm; 73 $\mu W/cm^2$) and a blue (470 nm; 115 $\mu W/cm^2$) LED. Additionally, steady state violet (400 nm, 2800 $\mu W/cm^2$), red (624 nm, 652 $\mu W/cm^2$), green (524 nm, 580 $\mu W/cm^2$) and blue (470 nm, 680 $\mu W/cm^2$) illumination (optical bias) was superimposed from the front (pin1) and back (pin2) sides in LEDs driven at different current values. The spectral sensitivity was analyzed by applying different wavelength optical bias from the front and backsides of the device (see figure1). Under front irradiation, the spectral photocurrent without light bias control at -8V applied voltage and different frequencies (250 Hz-3500Hz) is shown in figure 2.

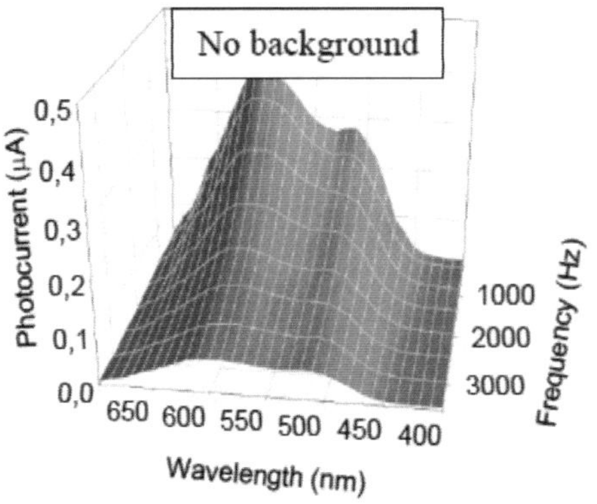

Figure 2. Spectral photocurrent without light bias control from the front side.

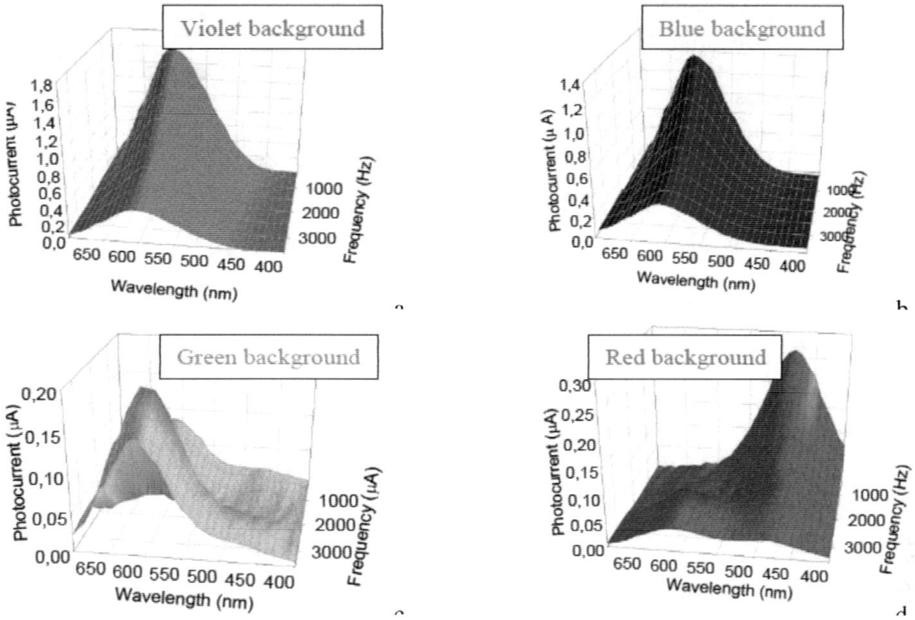

Figure 3. Spectral photocurrent under violet (a) blue (b) green (c) and red (d) bias control applied from the front side.

Under front irradiation, in figure 3 it is displayed the spectral photocurrent for different frequencies under violet (a), blue (b), green (c) and red (d) light bias control at -8V applied voltage and different frequencies (250 Hz-3500Hz). Results show that the spectral response depends strongly on the bias control wavelength and frequency. As the bias control wavelength increases the spectral sensitivity shifts to the low wavelength spectral regions and decreases with the frequency, suggesting capacitive effect across the device.

In figure 4 it is displayed the spectral photocurrent at 3500 Hz, under red, green, blue and violet background irradiations (colour symbols) and without it (black symbols) applied from the front (a) and back (b) diodes. For comparison the spectral photocurrent (right axis) for the front, p-i'-n and the back, p-i-n, photodiodes (dash lines) are superimposed.

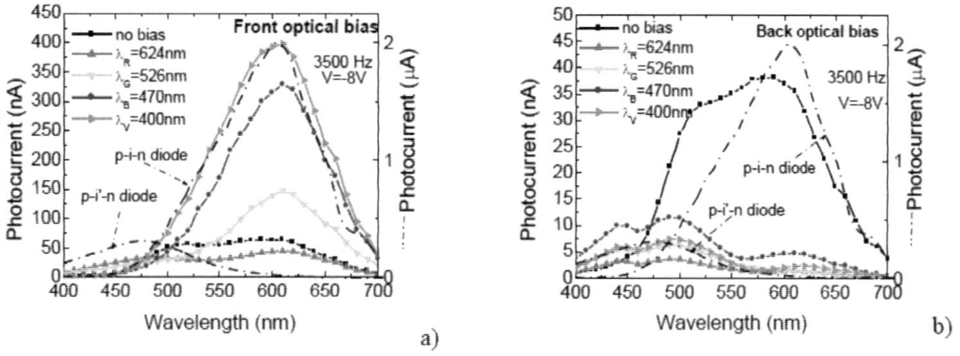

Figure 4. Spectral photocurrent under red, green, blue and violet background irradiations (color symbols) and without it (black symbols) applied from the front (a) and back (b) diodes.

Results show that the spectral sensitivity under steady state irradiation depends on its wavelength and on the impinging side. Under front irradiation, the back diode photoresponse is tuned and the sensitivity strongly increases for wavelengths higher than 500 nm when compared with its value without optical bias. As the background wavelength decreases, the spectral response increases. Under back irradiation the front diode photoresponse is selected. The sensitivity strongly increases in the short wavelengths range and collapses in the long wavelength region.

3. Transfer characteristics

In figure 5, at 3500 Hz, it is displayed the ratio between the photocurrent under different optical bias and without it (gain) under front (symbols) and back (lines) irradiations: red (α^R), green (α^G), blue (α^B) and violet (α^V).

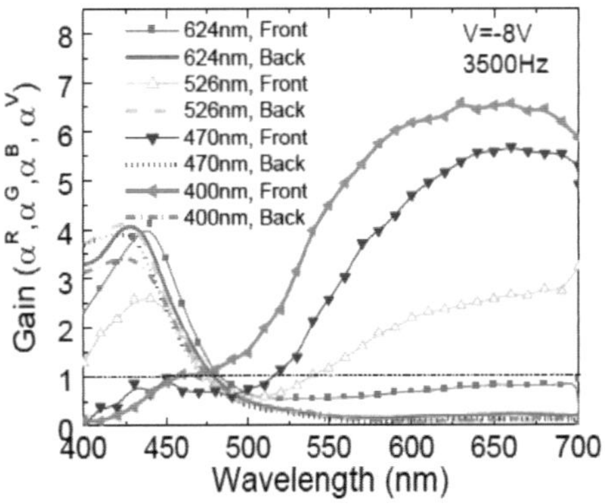

Figure 5. Gain under front (symbols) and back (lines) irradiations: red (α^R), green(α^G), blue (α^B), violet (α^V).

Under back irradiation (lines) the gain does not depend on the wavelength of the background. The spectral sensitivity in the low wavelength range is enhanced and the device always acts as a short-pass filter. Under front irradiation (symbols) the filter properties of the device depend on the background wavelength. Under red irradiation (figure 3d) the gain is high in the short wavelengths range acting the device as a short-pass filter. Under violet (figure 3a) and blue (figure 3b) irradiations, the transfer function presents an enhanced gain in the long wavelength range acting as a long-pass filter. Under front green background, the device (figure 3c) is a band-rejection active filter for frequencies higher than 2000Hz that works to screen out wavelengths that are within the medium range (475nm-550nm) giving easy passage at all wavelengths below and above.

Results confirm that under controlled wavelength backgrounds it is possible to fine-tune the spectral sensitivity of the device. Its sensitivity is strongly enhanced (α>1) in a specific wavelength range and quenched (α<1) in the others, tuning or suppressing a specific band. The sensor is a wavelength current-controlled device [10] that makes use of changes in the wavelength and impinging side of the optical bias to control the power delivered to the load. Self-optical bias amplification or quenching under uniform irradiation is achieved. By using background lights with complementary light penetration depths across the pi´n/pin device and changing the irradiation side, it is possible to control the spectral response and to filter a specific wavelength band.

4. Photonic active filters

To analyze the device under information-modulated wave and uniform irradiation, three monochromatic pulsed lights separately (red, green and blue input channels) or their combination (MUX signal) illuminated the device. Steady state red (652μ Wcm^{-2}), blue (680μ Wcm^{-2}) and violet (2800μ Wcm^{-2}), optical bias was superimposed separately from the front (pin1) and the back (pin2) sides and the photocurrent measured at -8 V. In Table 1 the gains (α) of the individual channels, defined as the ratio between photocurrents under irradiation (ON state) and without it (OFF state), for the red, green, and blue input channels are presented. Here, the superscripts are related to the background wavelength (R, G, B, V) and the subscripts (Rpin1,2,G pin1,2 B pin1,2) to the channel color and irradiation side.

Table 1. Gains (αR,G.B,pin1,2) at the input red, green, and blue channels wavelengths

Channels	α^R	α^G	α^B	α^V
$\alpha_{R,pin1}$	0,62	1,28	3,00	5,13
$\alpha_{R,pin2}$	0,12	0,42	0,47	0,34
$\alpha_{G,pin1}$	0,63	0,85	1,52	2,40
$\alpha_{G,pin2}$	0,59	0,62	0,67	0,67
$\alpha_{B,pin1}$	1,39	1,00	0,81	1,11
$\alpha_{B,pin2}$	2,22	1,94	1,97	1,64
$\alpha_{V,pin1}$	11,57	5,80	1,60	1,00
$\alpha_{V,pin2}$	11,57	13,90	1,20	10,42

Results show that, even under transient conditions, the effect of the background wavelength and impinging side presents the same nonlinear dependence as in figure 5. The morphology of the filter results from the interaction of the electric field under applied optical bias (red, green, blue, violet) and the transient electric field induced by the input channels (red, green, blue and violet). Under back irradiation, the small absorption depth of the violet photons across the back diode quenches the electric field there and as a consequence the red collection (Red ON) almost disappears (α^VR,pin2<<1). The blue channel is absorbed across the front diode where the electric field was enhanced resulting in an increase collection of the channel (α^V B,pin2>1). Since the green channel is absorbed across front and back diodes, its collection is balanced by the increase collection in the front diode and its reduction at the back (α^VG,pin2~1). The front violet background is absorbed at the surface of the front diode, increasing the electric field at the back diode, where the red and part of the green channels generate optical carriers. So, the collection is strongly enhanced (α^VR,pin1>>1, α^VG,pin1>1) while the blue collection stays near its dark value (α^VB,pin1~1).

The polychromatic combinations of the same red, green, blue and violet input channels with gains presented in Table 1 was used to generate a multiplexed (MUX) signal.

In figure 6, the filtered signals under front (pin1) and back (pin2) violet light control are displayed. The signals were normalized to the maximum intensity under violet front irradiations to suppress the dependence on sensor and LEDs positioning. The bit sequences used to transmit the information are shown at the top of the figures.

Different gains for the RGB channels were obtained (Table 1). Due to this wavelength non-linearity under front violet background, the encoded multiplexed signal presents as many levels as the possible RGB combinations, in a maximum of 2^3(eight-level encode). Those levels can be grouped into two main classes due to the high amplification of the red channel (α^VR,pin1>>1). The upper levels are ascribed to the presence of the red channel and the lower to its absence allowing the red channel decoder. Since under front irradiation the green channel is amplified (α^VG,pin1>1), the highest levels, in both classes, are ascribed to the presence of the green channel and the lower ones to its lack (long-pass filter). Under back irradiation the red channel is suppressed (α^VR,pin2<<1), the blue enhanced (α^VB,pin2>1) and the green reduced(α^VG,pin2<1), so the encoded multiplexed signal presents a maximum of four separate levels (2^2). The highest levels correspond to the presence of the blue channel ON with or without the green ON respectively, and the other to its absence. The blue channel is then decoded using this simple algorithm (short-pass filter).

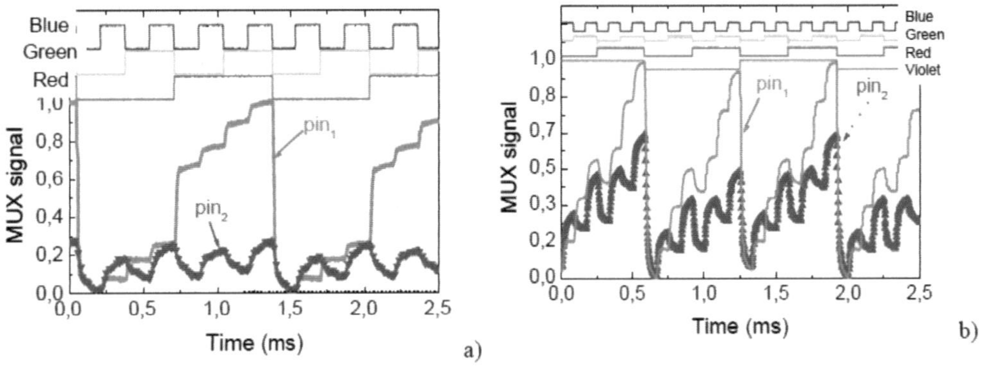

Figure 6. Filtered output signals under front (pin1; lines) and back (pin2; symbols) violet irradiation. a) Red, Green and Blue channels. b) Red, Green, Blue and Violet channels. On the top of the figures the optical signals used to transmit the information guide the eyes.

To test the algorithm, a polychromatic combination of the same red, green and blue input channels of figure 6a, in a different bit sequence, was used to generate a multiplexed (MUX) signal. In figure 7 the filtered signals under front (pin1) and back (pin2) violet light control are displayed. The signals were normalized to the maximum intensity under violet front irradiations to suppress the dependence on sensor and LEDs positioning. The decoded signal is shown on the top of the figure showing the effectiveness of the simple algorithm. Results show, that by means of violet optical bias control, the MUX photonic function may be modified giving reconfiguration. This new method based on wavelength background processing techniques enables the optical routing.

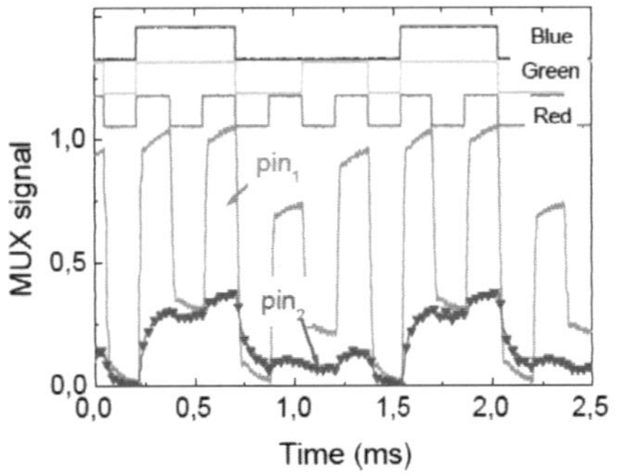

Figure 7. Filtered output signals: front (pin1; lines) and back (pin2; symbols) violet irradiation. On the top, the decoded signal is displayed.

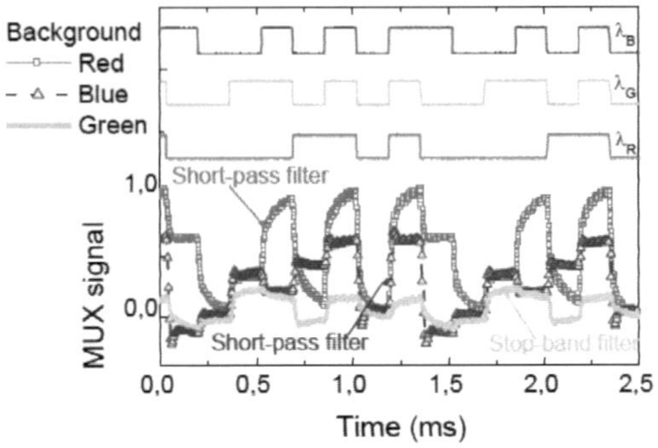

Figure 8. Filtered output signals under front red, green and blue background. The bit sequences used to transmit the information are shown at the top of the figures.

5. Optical bias controlled filtering

In figure 8, the filtered signals under red and green blue light control are displayed with the colour lines ascribed to the front background wavelength. The signals were normalized to the maximum intensity under red front irradiations. The bit sequences used to transmit the information are shown at the top of the figure.

Taking into account table 1, under red background, the red channel response is reduced ($\alpha^R R<1$) and the blue ($\alpha^R B>1$) and green ($\alpha^R G>1$) ones increased. The red channel is rejected (short-pass filter), so the mixture of both blue and green is recovered. The filtered signal is a 4-level encoding (2^2). Under blue irradiation, the optical bias effect is opposite and the blue channel is rejected (long-pass filter). So, the change in the transmitted information is due exclusively to the presence of both red and green channels. The filtered signal becomes a 4-level encoding (22^2). Under green irradiation, the devices act as a stop-band filter that drops the green channel. By subtracting the output signal under irradiation from the signal without it, the green channel is decoded. Here, the filtered information is mainly a 2^1 level encoded data.

6. Conclusions

Integrated photonic filters based on SiC multilayer devices were analyzed and their transfer functions presented, showing that the light-activated pi'n/pin a-SiC:H devices combine the demultiplexing operation with the simultaneous photodetection and self-amplification of an optical signal. Reconfiguration was provided by background wavelength control. Results show that the background wavelength and irradiation side control the output signal. By using background lights with complementary light penetration depths and changing the irradiation side, it is possible to change the spectral response and to filter a specific wavelength band acting either as a short- or a long- pass band filter or as a band-stop filter.

Acknowledgements

This work was supported by PTDC/EEA-ELC/111854/2009, PTDC/EEA-ELC/115577/2009 and PTDC/EEA-ELC/120539/2010.

7. References

[1] Connelly M J 2002 *Semiconductor Optical Amplifiers (*Boston, MA: Springer-Verlag) ISBN 978-0-7923-7657-6

[2] Petit C and Blaser M 2006 Photodiodes with integrated optical filters for passive optical network applications *Workshop on Optical Components for Broadband Communication* (ed. by Pierre-Yves Fonjallaz, Thomas P. Pearsall) Proc. of SPIE Vol. 6350 63500I

[3] Ibrahim S, Luo L W, Djordjevic S S, Poitras C B, Zhou I, Fontaine N K, Guan B, Ding Z, Okamoto K, Lipson M and Yoo S J B 2010 Fully reconfigurable silicon photonic lattice filters with four cascaded unit cells paper OWJ5. *Optical Fiber Communications Conference* OSA/OFC/NFOEC (San Diego, 21 Mar 2010)

[4] Vieira M, Louro P, Fernandes M, Vieira M A, Fantoni A and Costa J 2011, "Three Transducers Embedded into One Single SiC Photodetector: LSP Direct Image Sensor, Optical Amplifier and Demux Device",*Advances in Photodiodes*, Gian Franco Dalla Betta (Ed.), ISBN: 978-953-307-163-3 (InTech Chap 19) pp:403-425

[5] Louro P, Vieira M, Vieira M A, Fernandes M and Costa J 2011 Use of a SiC:H Photodiodes in Optical Communications Applications *Advances in Photodiodes*, Gian Franco Dalla Betta (Ed.), ISBN: 978-953-307-163-3 (InTech, Chap 19) pp 377-402)

[6] Igushi Y and Yamabayashi N 2000 Novel rear-illuminated 1.55 μm-photodiode with high wavelength selectivity designed for bi-directional optical transceiver *Proc. 2th Int. Conf. on InP and Related Mater* vol 317 pp 317-320.

[7] Vieira M A, Vieira M, Costa J, Louro P, Fernandes M and Fantoni A 2011 Double Pin Photodiodes with Two Optical Gate Connections for Light Triggering: A Capacitive Two-phototransistor Model *Sensors & Transducers Journal* **10** Special Issue 96-120. ISSN 1726-5479.

3-D LTCC microfluidic device as a tool for studying nanoprecipitation

J N Schianti[1,3], N P N Cerize[2], A M Oliveira [2], S Derenzo [2] and M R Góngora-Rubio [2]
[1]University of São Paulo, Integrated System Laboratory, Polytechnic School.
[2]Institute of Technological Researches of São Paulo, Center of Processes and Products Technology. 05508-901, São Paulo, São Paulo, Brazil

E mail: junovais@ipt.br and julia.schianti@usp.br

Abstract. Nanoparticles have been used to improve the properties of many cosmetic products, mainly the sunscreens materials using nanoencapsulation or nanosuspensions, improving the contact with active molecules, enhancing the sun protection effect and facilitating formulations in industrial products. Microfluidic devices offer an important possibility in producing nanoparticles in a simple way, in one step bottom up technique, continuum process with low polidispersivity, low consumption of reagents and additives. In this work, we microfabricated a 3-D LTCC microfluidic device to study the nanoprecipitation of Benzophenone-3, used as a sunscreen in pharmaceutical products. It was observed that some parameters influence the particle size related to the total fluid flow on device, the ratio between phases, and the Benzophenone-3 initial concentration. The influence of applied voltages on particle sizes was tested also. For the processing, a high voltage was applied in a Kovar tube inserted in the 3D device. The use of microfluidic device resulted in particles with 100 up to 800 nm of size, with polispersivity index below 0.3 and offering an interesting way to obtain nanoparticles. These studies are still ongoing, but early results indicate the possibility of obtaining B-3 nanostructured material.

1. Introduction

Microfluidic devices could be microfabricated in different substrates such as glasses, silicon, polymers and ceramics, each of them with many potential advantages depending on the application. The alumina-based ceramic known as LTCC (*Low Temperature Co-Fired Ceramic*) offers the possibility to microfabricate complex 3D structures using a multilayer process [1, 2, 3].

The anti-solvent precipitation is one of the technique used to obtain nanoparticles and can be improved by the utilization of microfluidic devices [4, 5]. Using microfluidic devices it is possible to obtain a high degree of supersaturation with a uniform spatial concentration distribution resulting in particles with smaller nucleation dimensions [6]. The microfluidic devices offers the possibility of nanoparticles produced in only one step, in a continuum process with low polidispersivity, low consumption of reagents and additives [7]. Many kinds of materials could be synthesized in microfluidic devices leading to nanoparticles, such as polymers, inorganic materials, hydrophobic drugs, etc. Usually, the microfluidic devices geometries used for nanoprecipitation have a planar

[3] To whom any correspondence should be addressed.

structure or 2D geometry. However, the use of 3D structures increases the mass transfer on the system that could result in particles with smallest sizes [8, 9].

Benzophenone-3 (oxybenzone - $C_{14}H_{12}O_3$) has been used in topical applications such as sunscreen in the cosmetics industry and have properties which are improved in the nanoscale [10]. In this work we present the results of a study of Benzophenone-3 nanoprecipitation using a 3-D LTCC microfluidic device. Some parameters were analyzed to explore the influence on the particle size of the total fluid flow on device, the ratio between phases, and the Benzophenone-3 initial concentration. The influence of applied high voltages in the device cross-junction on the particle sizes was also tested.

2. Materials and Methods

2.1 Microfluidic device fabrication

The microfluidic device was microfabricated in LTCC 951 (DuPont), using a CO_2 Laser (Versalaser, ULS) and standard lamination and sintering process. In the cross-junction, a Kovar (TAI) tube is inserted to obtain a 3D structure, as showed in Figure 1. The device is also fitted with two electrical conductors for High Voltage (HV) connect to the fluids in order to observe nanoparticle size modifications. This 3D device, with the schematic as showed in Figure 2, is operated in the flow focusing regime in order to carry out the nanocrystallization process.

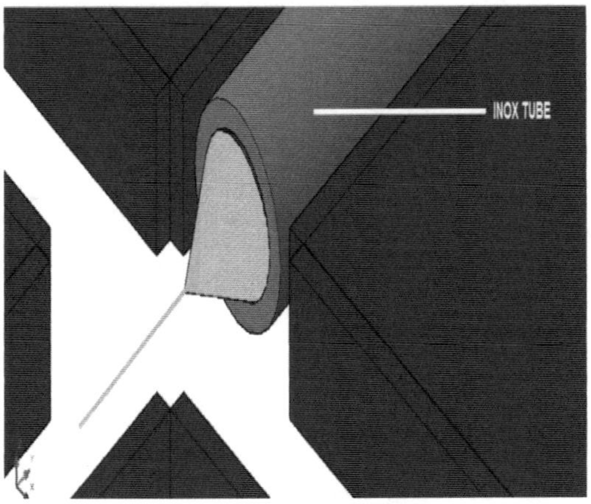

Figure 1. 3D flow focalization for nanoprecipitation process.

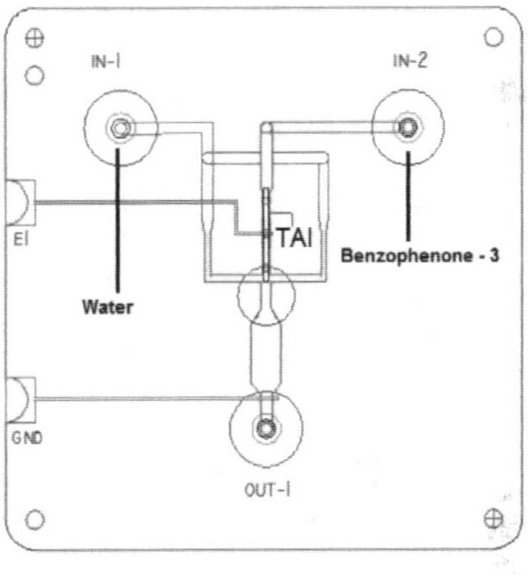

Figure 2. 3-D microfluidic device scheme for B-3 nanoprecipitation.

2.2 Nanoprecipitation Process

Two syringe pumps (Harvard Apparatus) were used to control the fluid flow, with pumping values ranging from 500 up to 1000 μL/min. Benzophenone 3 (B-3) was dissolved in two different solvents, isopropyl alcohol (IPA) and methanol (MTH), to observe the effect in the particle sizes. The influence of the total fluid flow rate was also observed, as well as the ratio between fluid flows ($R = Q_{water}/Q_{B-3}$) and the B-3 concentration. A high voltage source (GAMMA) with outputs ranging from 0 to 10KV was used to observe the influence of large electric fields in the nanoprecipitation. The B-3 nanoparticles morphology was analyzed by SEM-FEG and the particle sizes obtained by DLS (Delsa NANO).

3. Results

In Figure 3 we showed the graph of particle size versus the ratio between fluid flows ($R = Q_{water}/Q_{B-3}$) for two different initial concentrations of B-3 (2,5 and 7,5 mg/mL). In Figure 3a we can observe the result for the water flow rate (Q_{water}) of 1000 μL/min and in Figure 3b, Q_{water} of 500 μL/min. It is possible to observe that for higher particle concentrations we have bigger sizes. When we consider low fluid flow rates we have also particles with bigger sizes, so for this particular microfluidic devicethe best results are obtained for higher fluid flows rate and lower concentration. The ratio between phases R is a important parameter and for higher values of R we have particles with small sizes.

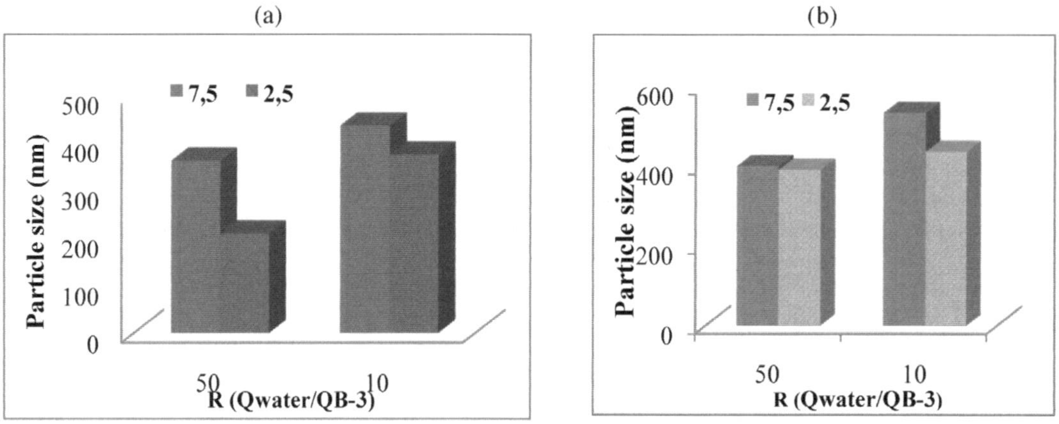

Figure 3. Graph of particle sizes versus R for two different concentrations. (a) considering Q_{water} = 1000 μL/min and (b) Q_{water} = 500 μL/min.

Figure 4 depicts the influence of the flows ratio on the particle sizes for two different solvents: Methanol and Isopropyl alcohol. It can be observed that for higher values of R is possible to obtain small particles. The size differences obtained for IPA and MTH were not particularly substantial. The MTH have more toxicity, so the use of IPA presents a better strategy for nanoprecipitation.

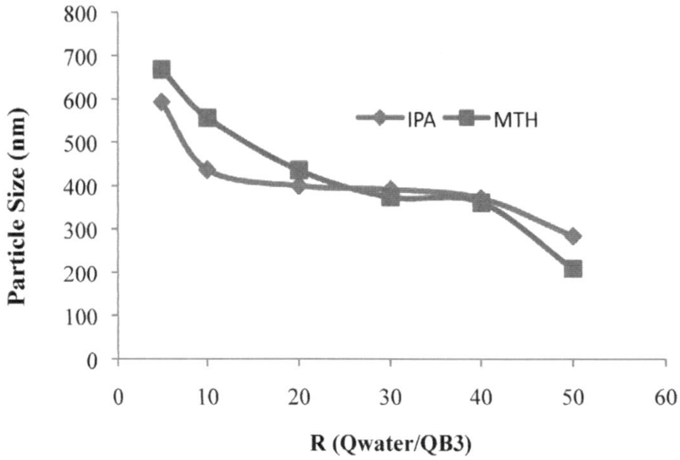

Figure 4. Particles sizes versus ratio between Water phase and B-3 phase using two different solvents: IPA and MTH.

Figure 5 shows the influence of applied voltage in the particle size for two different total fluid flow rates (500 and 1000μL/min). In these studies B-3 was dissolved in IPA with a concentration of 7,5 mg/mL. It was noted in both cases that the applied voltage increases the B-3 sizes. It was expected that the high voltage could decrease the particle sizes; however this phenomenon depends on many factors such as conductivity, inox tube size used on the microfabrication process, and the distance between electrodes, so the use of high voltages should be studied in more depth.

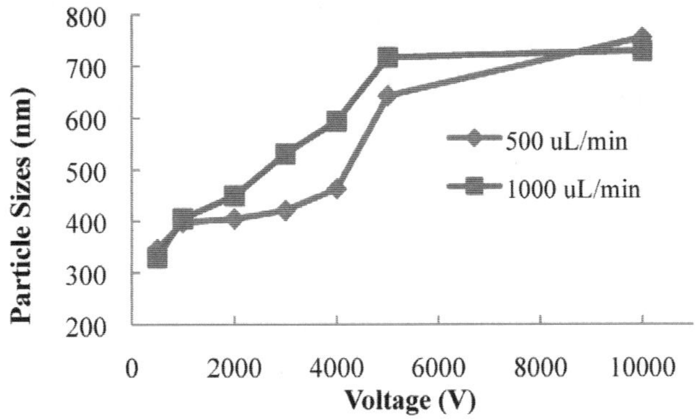

Figure 5. Particle sizes versus voltage for two different water fluid flows.

To observe the particle morphology a SEM-FEG analysis was done and the results are shown in Figures 6 and 7. This micrograph shows an example of particle size measured using DLS equipment. This example was obtained for a total fluid flow rate of 500 μL/min and 500 V. Note that the B-3 particles have spherical morphology and a size of 460 nm with low polidispersity.

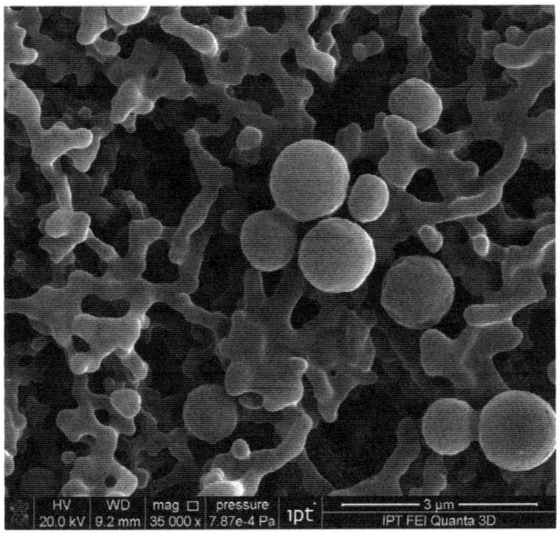

Figure 6. SEM-FEG analyses of B-3 nano-particles.

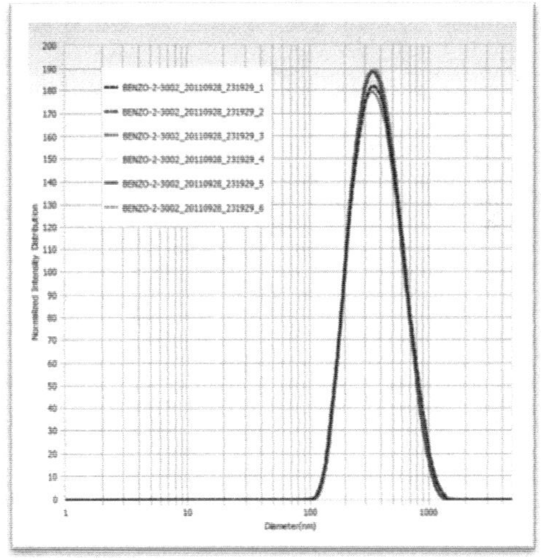

Figure 7. Particle size measured in DLS equipament.

4. Conclusions

Microfluidic devices are good tools in the production of nanoparticles. We introduce in this work a 3D microfabricated reactor in LTCC substrate. The use of this microfluidic device resulted in particles with diameters from 100 to 800 nm, with polispersivity index below 0.3 and offers an interesting way to manufacture nanoparticles. It is important to note, that in this microfluidic device the total flow rate is considered high when compared to a 2D structure and it is possible to work in the range of mL/min. The increased relation between phases (water and B-3) results in decreasing of particles dimensions. However, the use of high voltages results in larger particle sizes.

5. References

[1] Gongora-Rubio M R et. al. *Sens. Actuators A* 2001 **89** 222-24

[2] Ibanez-Garcia N et al. *Anal. Chem.* 2006 **78** 2985-92

[3] Ibanez-Garcia N et al. *Sens. Actuators B* 2006 **118** 67–72

[4] Panagiotou T et. al. *NSTI-Nanotech* 2007 ISBN 1420063766 4

[5] Marre S and Jensen K F *Chem. Soc. Rev* 2010 **39** 1183–1202

[6] Desportes S et. al. *Chemical Physics Letters* 2007 **446** 212–16

[7] C T Lo et. al. 2010 *Langmuir* **26** (11) 8559–66

[8] Davies T R et. al., *J. Micromech. Microeng* 2012 **22** 055003 (8pp)

[9] Rhee M et. al. *Adv. Mater* 2011 **23** H79–H83

[10] Samia D M et. al. *Current Nanoscience* 2010 **6** 654-65

Nanofibers and thin films as a selective membrane for sensors and microTAS

A N R da Silva[1,2,4], **M L P da Silva**[1,2] **and E R Fachini**[3]
[1]Department of Electronic Systems, University of Sao Paulo, São Paulo 05508-090, SP, Brazil
[2]Department of Electronic Systems, Faculty of Technology of Sao Paulo, São Paulo 01124-000, SP, Brazil
[3]College of General Studies, University of Puerto Rico at Río Piedras, San Juan 00931, PR, USA

Email: neilde@lsi.usp.br; malu@lsi.usp.br; erosim@gmail.com

Abstract. A composite material of PAN (polyacrylonitrile), starch granules and dimethylformamide as solvent was used as a selective membrane for volatile organic compounds (VOCs) in gaseous phase. This composite was produced as a thin film obtained by spin-casting and as a fiber mat produced by electrospinning. The fiber mat was tested for adsorption of VOCs and water. Characterization used microscopy (electron scanning and optical) in order to evaluate the fiber morphology and the starch incorporation in the PAN matrix. Infrared spectroscopy was intended to determine the starch presence. Relative viscosity of the starch/PAN suspensions was measured in order to provide a model of composite fiber formation. Quartz crystal microbalance experiments determined VOCs and water adsorption. Fibers incorporated the starch granules, making the composite sensible to water; but VOCs were not detected. Therefore, the composite was found to be a good choice as selective barrier on sensors or microTAS protection purposes.

1. Introduction

Electronic devices use membranes in several applications, such as protective layers [1] to electronic tongue or nose developments [2,3,4]. Composite membranes can merge different functions in one material, which is useful for micro-unit operations, such as microreactors, mixers, micro and nanofiltration, or as active parts on a microTAS (micro total analysis system)[5]. As an example, nanoparticles can be distributed in a porous polymeric matrix; in this case the disturbance in the flow through the porous improves mixing in microreactors. Another possibility is an online control of VOCs contamination in air or water flow using starch granules protected by a fluorinated film [6,7].

Due to the small dimension of electronic devices, the polymeric membranes must be thin (sub micrometer range) and able to cover a rough topography. On such approach, nanofibers are good candidates for protection/modification of such sensors and devices. The disturbance in flow caused by the nanofiber membrane will provide mixing and, eventually, removal of some contaminants. Additional advantages are low diffusion effects, which will not occur in conventional or in tunneling systems such as the one proposed by Macagnano [8].

Nano and microfibers can be produced by several different methods, with electrospinning being an

[4]To whom any correspondence should be addressed.

ease and cost-effective process [9,10]. Several research and technological applications use polymer fibers varying the compound and geometry of the fiber. The inherent large fiber surface is explored in sensors. The fibers can be modified in different ways, e.g. metals can be electroless-deposited or active particles can be incorporated into electrospun fibers. The incorporation of particles is a promising field of research in microreactor and pre-concentrators.

As far as we know, this work proposes for the first time the production of PAN/starch electrospun fibers. They were deposited on miniaturized structures for water detection.

2. Experimental

Fibers were produced by electrospinning using dispersion of starch granules from Cassava (*Manihot esculenta* Crantz, 5μm diameter, homemade produced) [6] and polyacrylonitrile (PAN, Sigma Aldrich) in N,N-dimethylformamide (DMF, Sigma Aldrich). PAN/starch mass ratio was: 1:0.1; 1:1; 1:3, 1:5 and 1:7. The electrospinning setup was described elsewhere [11]. To establish a model for starch incorporation, D-glucose (Casa Americana Ltda)/PAN in DMF was also electrospun. Films were obtained by spin-casting the same suspensions on silicon substrates at 1000 rpm, 6 minutes.

Fibers were characterized by optical (Leica Mycrosystem GMB) and scanning electron microscopy (SEM, Jeol JSM-6360); infrared spectroscopy (Thermo Nicolet Nexus 870) helped to determine the starch presence.

Adsorption of volatile organic compounds (VOCs) and water was verified using Quartz Crystal measurements (QCM) [6]. The relative viscosity of starch/PAN dispersions was measured, using an adapted Ford cup rheometer.

3. Results and Discussion

This narrative describes the experimental results in the following sequence:

- The film production and preliminary characterization;
- The establishment of a model for starch/fiber composite formation;
- The use in miniaturized structures.

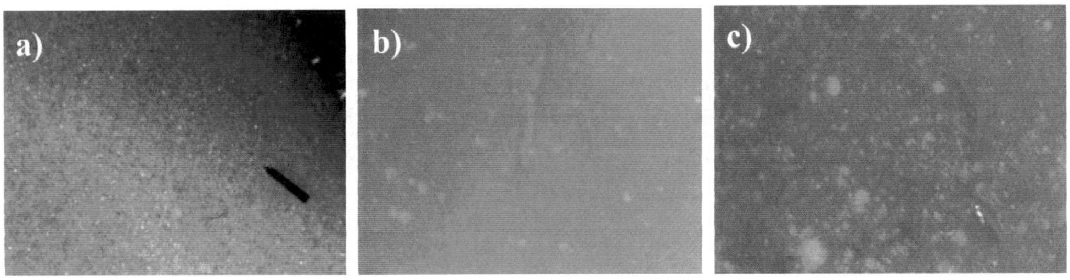

Figure 1. Optical image of films of a) starch; b) PAN; c) PAN:starch granules (1:1 w/w)

Figure 1 shows optical images of films produced by spinning process. The films appear homogeneous and similar. The PAN/starch sample presents more granules because starch is insoluble in DMF. Samples stocked for six months did not peel or change optical appearance, which pointed out the stability and low stress state of the PAN/starch composite.

Figure 2 shows optical microscopy for PAN/starch composite fibers in several mass ratios; it is noticeable that starch is incorporated into the fibers. Nonetheless, for mass ratio higher than 1:4, the starch granule coalescence is significant and, as can be seen in figure 2 (e), big clusters are detected. However, these clusters showed a surface quite different from the starch film of Figure 1 (a), suggesting strong interaction among nitrile groups from PAN and OH radicals from starch. Probably the PAN chains involve starch granules in the liquid dispersion, remaining attached even if precipitation of starch occurs.

a) PAN b) PAN/starch 1:1 c) PAN/starch 1:2

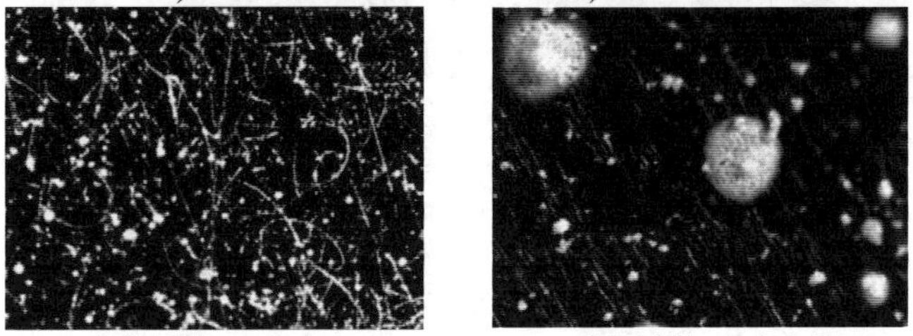

d) PAN/starch 1:4 e) PAN/starch 1:7

Figure 2. Optical microscopy for PAN/starch composite fibers with different mass proportions (w/w).

SEM images reinforce the assumption of strong interaction, probably hydrogen bond formation, between starch granules and PAN chains. As can be seen in Figure 3, the starch incorporation promotes the formation of nodules along the fiber. If low PAN/starch w/w ratios are employed, some starch grains could not be covered by the polymer chains, as the arrow in figure 3 (b) indicates. In figure 3 (b) the biggest granule shows a crack possibly due to charged particles desorption, suggesting the using of starch as an absorber inside the fiber [6].

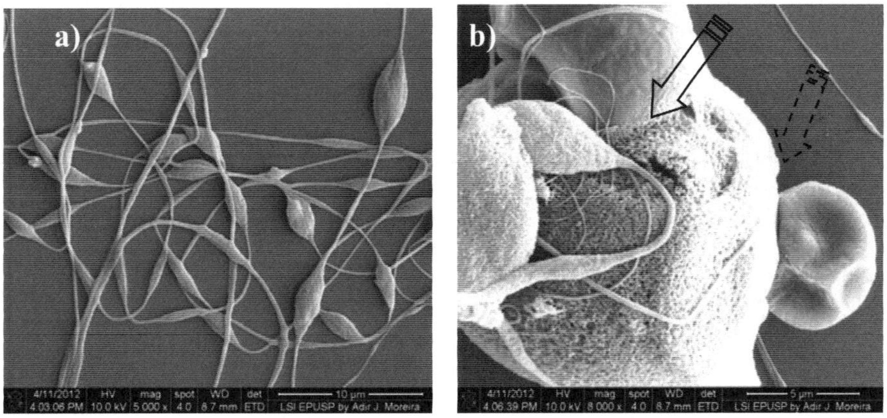

Figure 3. (a) The starch granules incorporated to the polymeric fibers. (b) The agglomerated granules covered by the polymer and an individual uncovered starch granule (dashed arrow). PAN/starch ratio (1:7)

The fiber morphology and diameter were strongly dependent of the amount of starch in the liquid dispersion as can be seen in Figure 4.

Figure 5 shows the fiber diameter as a function of the PAN/starch w/w ratio. The non linearity shown in the graph is consistent with two different regimes of starch incorporation, as pointed out by microscopy, and presents good agreement with a slope change at PAN/starch 1:3 mass ratio.

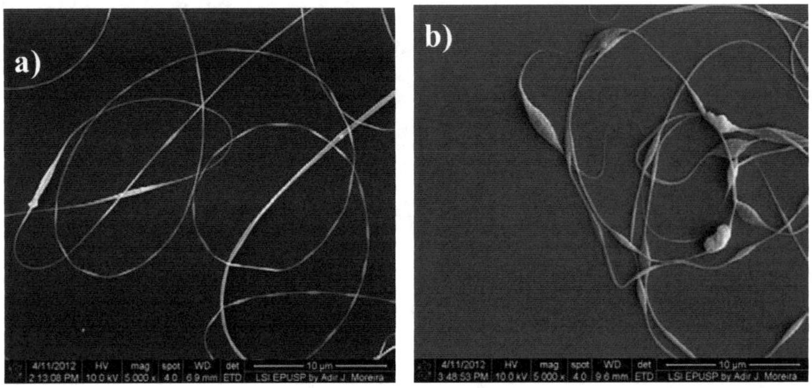

Figure 4. SEM images from nanofibers spun from dispersion with a) 1:1 and b) 1:5 w/w PAN/starch.

In order to get insight on starch/PAN chemical interaction, FTIR analysis was carried out; the FTIR results are shown in Figure 6. In this figure, bands for PAN fiber spectrum (continuous line) are consistent with the assignment made by Serkov [12]; bands are well defined, indicating a high ordered polymer structure. However, even for mass ratio of 1:0.1, the inclusion of starch granules broads the bands showing the interaction among fiber components. Bands assigned to CO (1190-950 cm^{-1}) and OH (3600-3000 cm^{-1}) species are characteristic of cassava starch [13]. The CH_n bands (2933 and 2881 cm^{-1}) are merged due to strong interaction of starch and PAN chains. The nitrile radicals' interaction no longer plays an important role since the band at 2245 cm^{-1} almost disappears and the band at 1590 cm^{-1} shifts to lower wavenumbers. In other words, hydrogen bonds with starch radicals predominate. These results agree with the SEM analysis. The increase in diameter is owing to particle insertion on the fiber and correlated with bead formation due to the strong interaction among PAN chains and starch.

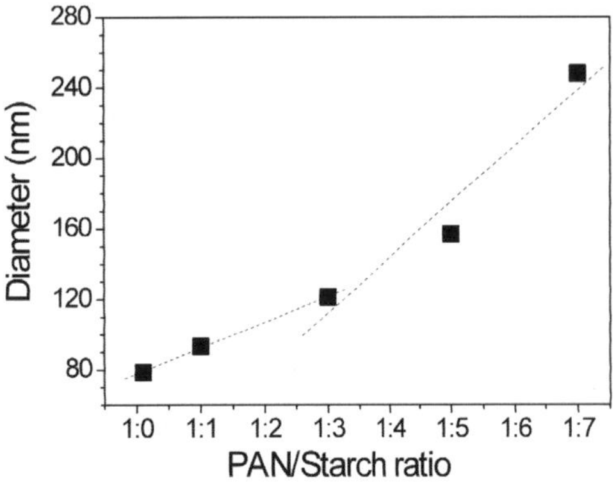

Figure 5. Fiber diameter variation in function of starch concentration.

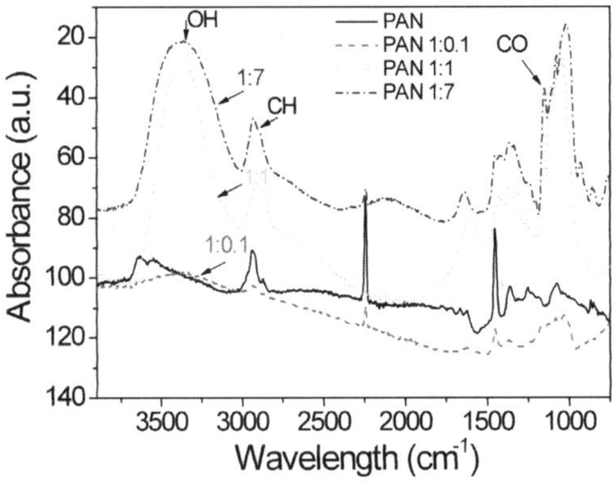

Figure 6. FTIR spectra for PAN and PAN/starch granules.

The two different regimes showed in figure 5 correlated with viscosity data. As can be seen in the figure 7, whereas the relative viscosity is almost constant in the range of 1:0 to 1:3 of PAN/starch w/w ratio, additional starch suddenly increases the viscosity. The detail on the graph shows a minimum on the relative viscosity at a ratio of 1:0.3. Therefore, up to a 1:0.3 ratio, the inclusion of starch granule in the dispersion decreases the relative viscosity probably due to the adsorption of the polymer molecules on the surface of the granule that might even precipitate. Thus, the system acts as if lower amounts of polymer were present in the liquid dispersion.

After the 1:0.3 ratio, additional starch enhances the interaction among granules and viscosity strengthens. When PAN/starch ratio overpasses 1:5, the starch and PAN in the dispersion act as a network; as a consequence, the viscosity increases suddenly. When that occurs, the fiber formation is difficult and higher fiber diameters will be achieved.

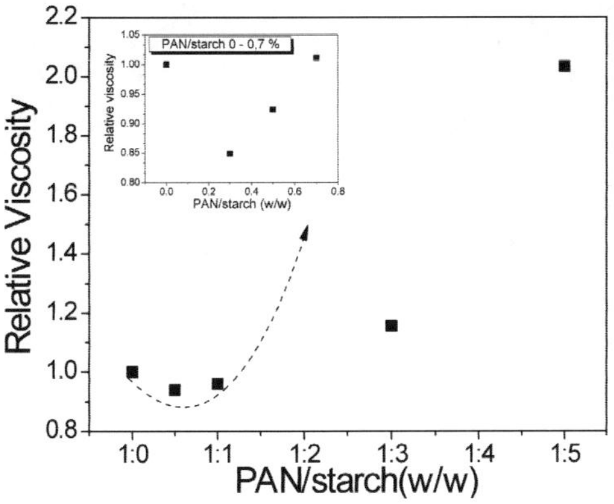

Figure 7. Relative viscosity in function of starch concentration.

This model for the PAN/OH-radical interaction was evaluated using D-glucose as a substitute for starch granules. D-glucose was selected because it is the monomer responsible for starch production and it also dissolves in DMF up to 0.7% w/w (determined experimentally), which will favour hydrogen bond formation on PAN dispersion.

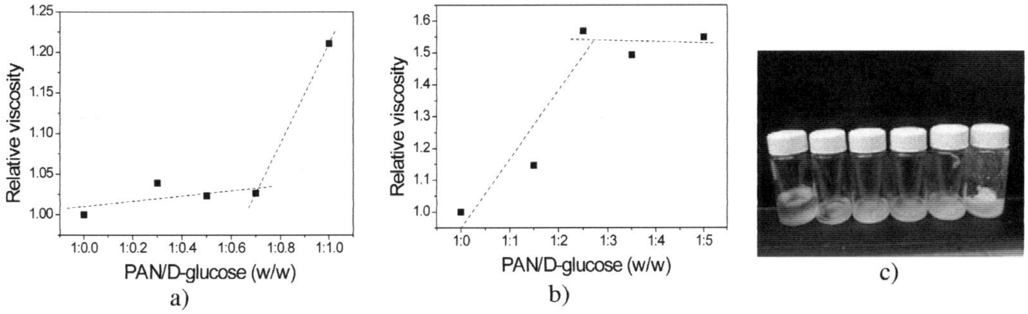

Figure 8. Relative viscosity as function of D-glucose concentration a) up to 1:1.0 (w/w), b) up to 1:5 (w/w). c) PAN/D-glucose dispersions. The solid phase is apparent on the right bottle.

Figure 8 shows typical results for relative viscosity as a function of D-glucose concentration. Up to 1:0.7 the 5% increase on viscosity is consistent with the existence of soluble D-glucose molecules. Additional D-glucose will not be solubilized, but the increase on relative viscosity will remain, i.e., up to the 1:2.5 range, addition of D-glucose promotes the formation of solid precipitates in a highly viscous medium. Furthermore, on the 1:3 region the strong interactions among these precipitates create a net, i.e., an almost solid material (Figure 8c), which hinders viscosity measurements.

SEM analysis (Figure 9) of PAN:D-glucose fibers shows rough surfaces and D-glucose crystal completely covered by the polymer, which points out strong interactions between PAN and D-glucose molecules in the whole material.

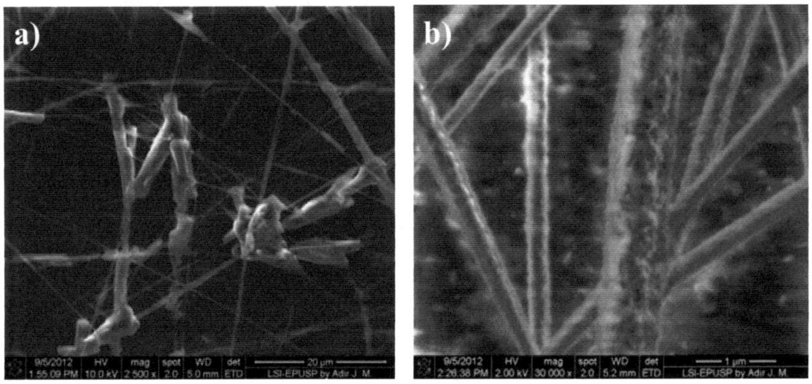

Figure 9. Fibers electrospun from dispersion of PAN and D-glucose. a) the D-glucose granules on the fibers surface at 2500 X and. b) at 30000 X, in this case it is possible to see the rough surface of the fiber due to the incorporation of the D-glucose.

In order to validate the use of PAN/starch fiber as an active material for detection, few fibers were electrospun directly on a) quartz crystals for QCM and b) in 3D-microchannels [6]. The fiber exposure to n-hexane or 2-propanol did not change the QCM signal; however, 10% of relative humidity caused a reversible change of the signal by 10 Hz. This means that water molecules could permeate the PAN chains (probably due to hydrogen bonds formation), while 2-propanol molecules could not, probably owing to the molecule size. Figure 10 shows typical results for QCM experiments.

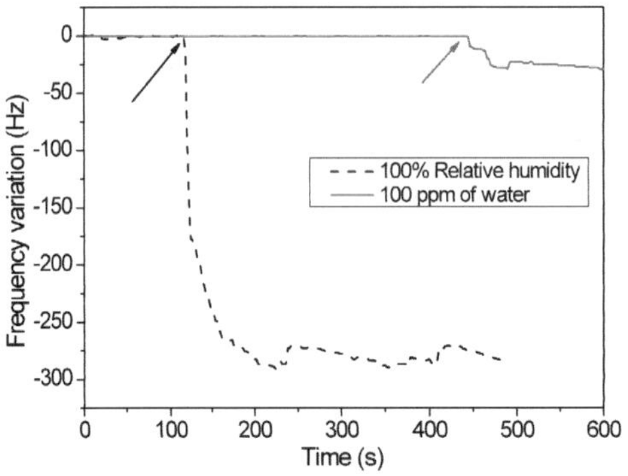

Figure 10. QCM measurements

A strong hydrogen bond linking nylon-6 nanofibers was already proposed [14], which reinforces the idea of water selective permeation from PAN fibers. Inside the fibers, starch will act as a scaffold and will retain a number of water molecules, which explains the 10Hz variation in the quartz oscillating frequency that means approximately 10% variation on the fiber weight. This phenomenon could be used in several ways, such as promoting preconcentration on microTAS or in a sensible layer for sensor applications.

The behavior of cassava starch granules was different on PAN/starch electrospun fibers than in a plasma polymeric thin composite film with fluor radicals obtained by Lima [6]. In fluorinated/starch composites, the permeable thin film allows the adsorption of 2-propanol. The use of fibers is more advantageous since it avoids diffusion effects that delay QCM measurements, such as the results found out using Zn-porphyrin film in similar devices [7].

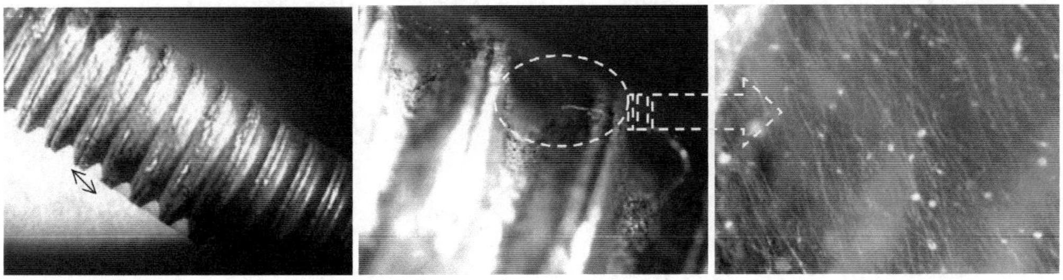

Figure 11. 3D-microchannels covered by PAN:starch fibers. Double arrow indicates 100 microns.

The ability of the PAN/starch fibers to deposit on 3D-microchannels is shown in the figure 11. Most of the fibers link the edges of each thread, forming a mat around the screw; however, some of these fibers were deposited in the channel, which is also interesting since it provides an easy way to insert adsorptive material in small spaces.

4. Conclusions
This work showed an easy way to produce PAN fibers modified with starch. Tests revealed that fibers completely cover the starch grains and a model was proposed to explain the experimental results. The produced composite fibers are sensible for water but not to VOCs, which means that these fibers can be a useful material for production of selective membranes. Electrospun fiber over the miniaturized structures showed a good coverage over the topography, which means that the fibers mat is a good choice as selective membrane for sensors or to protect microTAS surfaces.

Acknowledgments
To the Brazilian research funding agencies FAPESP and CNPq for financial support. A. J. Moreira for SEM analysis.

5. References

[1] Sherwood C, Harradine M A, Birch T S and Stevens J C 1996 *Microelectronic Eng.* **30** 547-550
[2] Riul Jr A, Gallardo Soto A M, Mello S V, Bone S, Taylor D M and Lattoso L H C 2003 *Synthetic Metals* **132** 109–16
[3] Vilan A and Cahen D 2002 *Trends in Biotechnology,* **20**, 1, 22-9
[4] Lu J G, Chang P and Fan Z 2006 *Materials Science and Engineering: R: Reports* **52**, 1-3, 49-91
[5] Ulbricht M 2006 *Polymer* **47** 2217–62
[6] Lima R R, Hernandez L F, Carvalho A T, Carvalho R A M and da Silva M L P 2009 *Sensors and Actuators B* **141** 349–60
[7] Elosua C, Matias I R, Bariain C and Arregui F J 2006 *Sensors* **6** 1440-65
[8] Macagnano A, Zampetti E, Pistillo BR, Pantalei S, Sgreccia E, Paolesse R and d'Agostino R 2008 *Thin Solid Films* **516**, 21, 7857-65
[9] Huang Z M, Zhang Y Z, Kotaki M and Ramakrishina S 2003 *Composites Science and Technology* **63**, 2223
[10] Nataraj S K, Yang K S and Aminabhavi T M 2012 *Progress in Polymer Science* **37**, 3, 487-513
[11] Gomes D S, Silva A N R da, Morimoto N I, Mendes L T F, Furlan R and Ramos I 2007 *Polímeros* **17** 206-11
[12] Serkov A T 2007 *Fibre Chemistry* **39**, 1, 60-3
[13] Kaewtatip K and Tanrattanakul V 2008 *Carbohydr. Polym.* **73,** 4, 647–55
[14] Pant H R, Bajgai M P, Yi C, Nirmala R, Nam K T, Baek W and Kim H Y 2010 *Colloids and Surfaces A: Physicochemical and Engineering Aspects* **370**, 1–3, 87–94

HSENSE: a high performance framework for distributed weather sensor networks

J A Ortiz[1], N G Santiago, J G Colom-Ustáriz
Department of Electrical and Computer Engineering, University of Puerto Rico, Mayagüez, Puerto Rico.

E-mails: jose.ortiz19@upr.edu, naydag.santiago@upr.edu, colom@ece.uprm.edu

Abstract. HSense is a framework designed for aggregating and monitoring weather sensor networks. It is implemented using the functional language Haskell to provide a modular and expandable interface and parallel data processing. HSense provides a building block in which researchers can integrate sensor networks and develop applications. It is specially developed for managing large quantities of distributed weather sensor data efficiently. This document summarizes the design and development of the HSense framework and a case study of the Puerto Rico Weather Radar Network at the University of Puerto Rico at Mayagüez.

1. Introduction

Distributed weather sensor networks are composed of multiple independent nodes used to measure weather parameters around their specific coverage area and at the same time maintain communication with each other. Vast data generated by these types of sensors is communicated, processed, and stored at common storing centers (called sinks) for further analysis. While several sensor network management methods exists, managing high quantities real-time data and develop applications for heterogeneous networks become a hurdle [1]. Most of weather sensor networks produce higher quantities of data and most applications are developed targeted to specific network architecture. However there is no framework for developing applications for high throughput generic sensor networks such as the one assumed in this article. The particularity of this problem of managing sensors like high performance weather radars entails the need of implementing a different alternative.

This work has the following objectives: first, to manage distributed weather sensor networks using high performance systems; second, to demonstrate the viability of using functional languages for managing sensors through the use of parallel processing. Finally, the third objective is to create a scalable and modular framework for managing weather sensors, capable of expanding and integrating heterogeneous systems.

Evaluations of current methods of managing sensor networks are presented and a proposal for an improved way of managing while adding capabilities of processing and storing sensor data. Modern approaches of manipulating data are implemented, several of these detailed in the subsequent sections. HSense framework introduces various emergent technologies into the field of distributed sensor networks (DSN). Web protocols are introduced to communicate between the

[1] To whom any correspondence should be addressed.

sensor nodes, distributed databases for chaining different database networks, and a functional language that provides built in high performance capabilities with high concurrency and highly modular design patterns.

A case study of the framework designed is provided. It uses the Puerto Rico Weather Network courtesy of the University of Puerto Rico at Mayagüez. As the Puerto Rico network contains several different weather sensors, it became an ideal candidate for experimenting with alternative ways of managing sensors. Moreover, due to the large quantities of data that maintains the network, several highly parallel algorithms and distributed databases are tested. The ability to aggregate data from heterogeneous weather sensor networks will create a whole new range of applications involving weather data and capable of complementing current weather technology prediction systems.

2. The HSense Framework

The HSense framework design focus aims to offer a set of libraries for managing and processing weather sensor networks. It is intended to provide a set of organized common steps that a researcher can follow to manage their own sensors and share its data with web users. The HSense framework has the following features:

- *Sensor Management* – capable of maintaining status of current connected sensors to the network with a notification system in case of sensor failure.
- *Parallel Processing Capability* – built-in support for highly parallel processing algorithms.
- *Basic Sensor Data Visualization* – provide tools for basic sensor visualization such as plots and graphics.
- *Web Interface* – a web-based interface to access the framework resources.
- *Standalone Interface* – an interface that allows researchers to use the HSense basic algorithms to manage sensor and process their data.
- *API Interface* – an Application Programming Interface to allow developers to take advantage of framework sensor data.

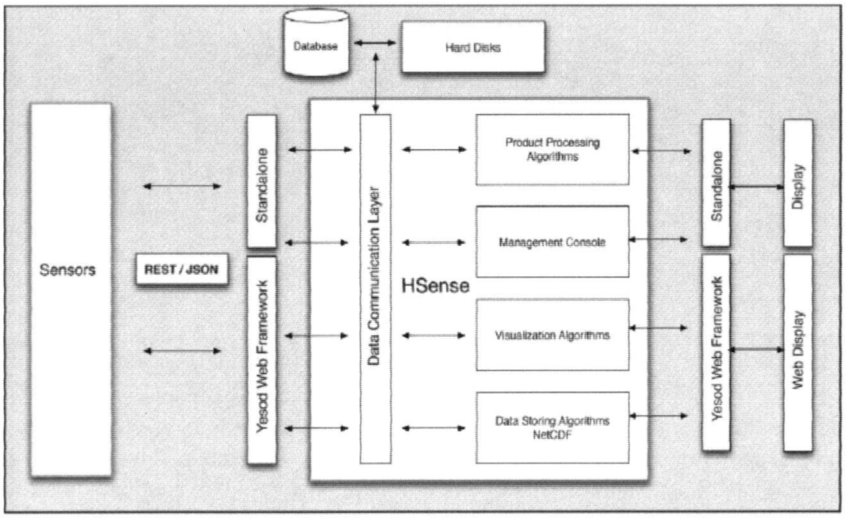

Figure 1: HSense Internals – Block Diagram of HSense framework and its interfaces

HSense is built using the Haskell functional programming language. Haskell provides built-in features like high concurrency and parallel capabilities while improving the code maintenance through modules. HSense contains a Haskell-based concurrent server with several modules for task division. As shown in the figure 1, each module manages their task independently while the main communication modules assign the tasks. The following sections describe the independent modules.

2.1. *Sensor Management Module*

The sensor management module permits real time monitoring of all the connected sensors. This allows the system to make decisions based on the sensor status. Moreover, it lets the network owner to stop sensors at will or validate if the sensor needs maintenance. Sensor management is a complex task by itself; HSense provides basic information and statistics for sensors and networks connected to the system. A simple knowledge engine will provide notifications to researchers about the sensors' states. The console is capable of being accessed online and manipulates features of the HSense based systems.

2.2. *Communications Module*

The communication module (see figure 2) is based on the web technologies REST and JSON. Representational State Transfer (REST) is a style of software architecture, which simplifies the way of creating web services [2]. It allows simple HTTP address to require data or push data; this simplicity permits the integration of any web server as a possible node of a network. The JSON protocol, based on the Javascript programming language, is a common vessel for transporting information through the Internet. Contrary to XML (previous common protocol for transporting data on web), JSON protocol is easier for human reading and simpler; therefore it has become standard in current Internet companies. JSON packets can travel inside firewalls and out to the Internet thanks to the use of port 80 for communication; this provides an advantage for the integration of intranet sensors.

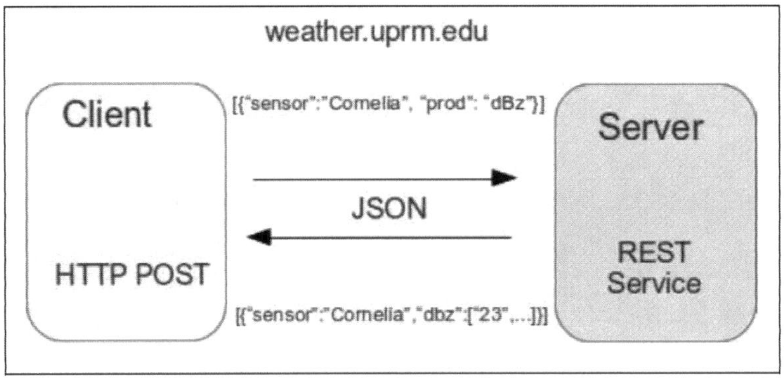

Figure 2. Communication – JSON based communication between sensors.

2.3 *Processing Algorithms*

The "PROCESSING ALGORITHM MODULE" contains algorithms for manipulating data. This data received from the sensors is processed in parallel and the results are stored in a database. This module also contains templates to allow the user to add algorithms to the platform. Adding an algorithm is as simple as dropping the code in a folder and the system will recompile and add it to the other available algorithms. Some processing algorithms that are being demonstrated are reflectivity calculations for rain, rain rate estimation, and plan position indicators (PPI) from weather radars.

These algorithms can be applied to large datasets of information from the database. This allows the user to generate reports or create studies based on the dates and times, which is critical when analyzing weather events. All parallel algorithms should be applied to data parallel structures (meaning arrays or arrays of arrays); this will ensure the most efficient parallelization. To ensure compatibility, all algorithms will use NetCDF file format as data entity [3]. The

NetCDF bindings created especially for HSense take advantages of the Haskell language for parallel algorithms.

2.4. Visualization Module

The "VISUALIZATION MODULE" generates images based on processed data to helps researchers analyze the data visually. Visuals are based on, if available, previously processed data. Several web services will allow the retrieval of previous generated visual information. The visualization module also will provide data as required by the web server module. The plots in Figure 3 show a common visualization of a weather event.

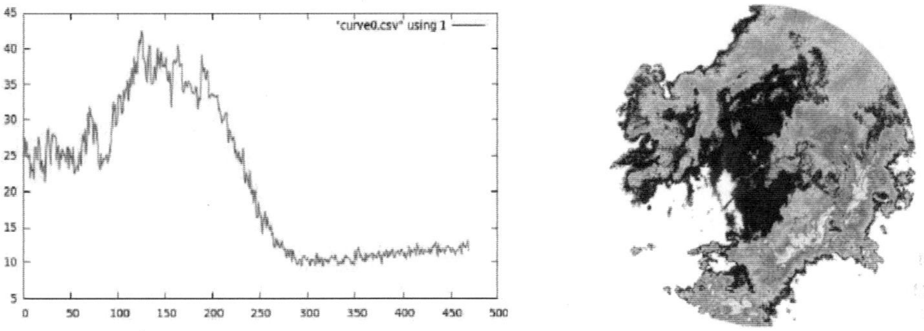

Figure 3. Weather Event Visualization – A plot of a beam of a weather radar and a PPI plot of the weather event

2.5. *HSense Interfaces*

The frameworks provide three interfaces: a standalone interface, an API interface, and a web interface. The standalone interface allows researchers to use HSense as a library for manipulating and visualizing weather data. It provides features for opening and saving NetCDF files, visualize radar data, and compare information from a Matlab-like console environment. The API interface provides developers the ability to build applications based on the information contained in HSense. The web interface serves as a template application for using the HSense API. It uses most of the API calls available for the developers and delivers an interactive website to allow users visualize real time weather sensor data.

Figure 4. Weather Portal - A website created using the HSense Framework for the PRWRN.

3. Puerto Rico Weather Radar Network: A Case Study

The Tropinet MRI and the CASA-ERC projects at the University of Puerto Rico Mayagüez (UPRM) were founded to create multiple weather radar networks in the west coast of Puerto Rico.

The CASA ERC Puerto Rico Testbed is a network of Off-the-Grid (OTG) weather radars to research lower atmosphere weather events [4]. The OTG is modified marine radar made capable to detect rain reflectivity. This CASA-ERC project aims for the development of low cost, low infrastructure weather radars capable of working with green energy.

The Tropinet MRI weather radar network provides state of the art radars to Puerto Rico west coast. The Tropinet weather radars are the only dual polarization Doppler weather radars in Puerto Rico [5]. Figure 4 shows the Tropinet Radar Network. Both networks provide important information to researchers about the weather conditions of the west coast of Puerto Rico. However, these networks lack a framework to integrate and manage their radar sensors. Given the development status of the Puerto Rico Weather Radar Network (PRWRN) was the perfect subject to test the capabilities of HSense managing weather sensors.

Based on the HSense framework API interface, we built a website for displaying current sensor data to users. It also permits researchers to access raw data for further analysis if needed, along with common visualization algorithms built for HSense. The Puerto Rico Weather Radar Network comprehends a distributed sensor network capable of integrate seamlessly using web technologies to any web based network. Due to the HSense framework, any other web capable network could share and connect to the sensors to create larger and more powerful networks.

4. Conclusions

In this article a design and development of HSense framework for managing distributed sensor networks is presented. We specified a design that used multiple technologies such as REST services, HTML5, and functional languages. This gave us the ability of developing a high performance framework for managing weather sensor networks. Functional languages such as Haskell helped in the development, incorporating parallelization, modularity, and concurrency. It was also showed parallel algorithms for processing weather data, NetCDF algorithms for data storage and visualization algorithms.

The HSense framework provides an Application Programming Interface (API) that permits the user to create applications that use the data of managed sensors. A template for managing sensors is provided which uses the HSense framework with an integrated web server for display sensor data. Finally a case study was developed using the template provided for managing the weather sensors of the Puerto Rico Weather Sensor Network, in which the capabilities of the framework were shown. Several tests were performed that showed that the platform is stable and capable of managing hundreds of sensors simultaneously.

There are several areas for the framework to improve. For example, it will be interesting to do testing with multiple remote sensor networks and with several databases. HSense provides an approach of management unique to high throughput weather sensors that will help researchers develop their own sensor networks, manage them, and make decisions based on the data stored.

5. References

[1] Mottola L and Pietro G 2011 Programming wireless sensor networks: Fundamental concepts and state of the art. *ACM Comput. Surv.* **43(3)** 19:1-19:51

[2] He J, Zhang Y, Huang G and Cao J 2012 A smart web service based on the context of things *ACM Trans. Internet Technol.* **11(3)** 13:1-13:23

[3] Rew R and Davis G 1990 NetCDF: an interface for scientic data access IEEE *Comput. Graph. Appl.* **10(4)** 76 82

[4] Trabal J M, Colom-Ustariz J G, Cruz-Pol S L, Pablos-Vega G A, and McLaughlin D J 2012 Remote Sensing of Weather Hazards Using a Low-Cost and Minimal Infrastructure Off-the-Grid Weather Radar Network *IEEE Trans. Geosci. Remote Sens.* **99** 1-15

[5] Trabal J M, Colom-Ustariz, Pablos-Vega J G, Ortiz J, Castellanos W, Cruz-Pol S, Leon L and Rodriguez-Solis R 2011 Low cost and minimal infrastructure Off-the-Grid XBand radar net-work development for the west coast of Puerto Rico *IEEE Radar Conference (RADAR)* pp. 1026 -1031

Portable measurement system for FET type microsensors based on PSoC microcontroller

D Garnier Fernández[1], A Blanco[2], A Durán[2], C Jiménez-Jorquera[3] and O Arias-de Fuentes[2,4]

[1] Centro de Investigaciones Médico-Quirúrgicas (CIMEQ), Calle 216 esquina 13, Siboney, Apto. 6096, La Habana, Cuba

[2] Instituto de Ciencia y Tecnología de Materiales (IMRE), Zapata y G, Vedado CP 10400, La Habana, Cuba

[3] Instituto de Microelectrónica de Barcelona (IMB-CNM), CSIC, Campus-UAB, 08193 Bellaterra, Barcelona, España

E-mail: oarias@imre.oc.uh.cu, oarias@fisica.uh.cu

Abstract. FET based microsensors have reached increasing importance in biomedical and environmental applications, due to their advantages over conventional sensors. These advantages are related with their small size, robustness, low cost, integrability, fast response time, and the small sample volumes required for their use. In this work, it is presented the design and implementation of a portable instrumentation system based on the CY8C29466 (PSoC) microcontroller for the measurement of ISFETs and CHEMFETs type microsensors in liquid samples. The system shows the measurement's data on a display, also allowing storage of data for further analysis by coupling the system to a PC serial port.

1. Introduction

The use of microsensors for monitoring of environmental parameters and clinical diagnosis is gaining interest due to their advantages over conventional sensors. Among microsensors those based on semiconductor technology offer additional advantages such as small size, robustness, low output impedance and rapid response [1, 2]. Besides, the technology used allows integration of circuitry and multiple sensors in the same substrate and accordingly they can be implemented in compact probes for particular applications e.g., *in-situ* monitoring and/or on-line measurements. In this field, Ion Selective Field Effect Transistors (ISFETs) and Chemically Modified Field Effect Transistors (CHEMFETs) have a special interest. They are particularly helpful for measuring pH and other ions in small volumes and they can be integrated in compact flow cells for continuous measurements. They are also important in the field of diagnosis and clinical applications, where it is required to have equipments able to carry out measurements of diverse variables in complex samples like blood and urine [3, 4].

Microsensors based on FETs require a conditioning circuit that polarizes them appropriately [5]. The systems developed for data processing and visualization of the acquired signal usually include a data acquisition card (DAC) associated to a personal computer (PC) but it is also possible to use microcontrollers and virtual instrumentation to achieve this goal [5-7]. In that sense, systems based on Programmable System on Chip (PSoC) microcontroller allow the integration of certain analogical

[4] To whom any correspondence should be addressed.

Published under licence by IOP Publishing Ltd

sections in the system's digital design. This fact notably reduces the number of components and connections on the board circuit. The use of PSoC also reduces the implementation time of prototypes of complex systems [5] offering opportunities for designing very low cost systems for sensor applications. Among these, Programmable System on Chip (PSoC) by Cypress MicroSystems has a wide range of analog and digital integrated circuits in a single chip that allows designs with good performance, autonomy and flexibility.

The goal of this work is the design and implementation of a portable system, based on a PSoC microcontroller, associated to FET microsensors. The system is able to carry out measurements and the data can be shown in the liquid crystal display (LCD) of the portable system and alternatively can be stored in the PSoC, allowing its later recovery and analysis in the laboratory through the coupling to a PC.

2. Materials and Methods

2.1. ISFET Sensors

ISFETs are defined from the 1970's as an adaptation of metal-oxide-semiconductor field-effect transistor (MOSFET). Such an adaptation consists on the substitution of the metal gate (G) by an ion sensitive membrane, an electrolytic solution and a reference electrode [1, 4, 5]. Figure 1 shows an ISFET schematic diagram.

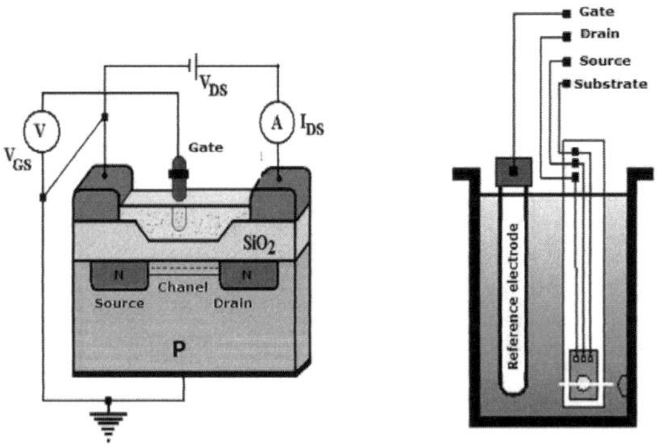

Figure 1. Schematic diagram of an ISFET.

The ISFET similarity with a MOSFET made possible its integration in electronic systems. As a chemical sensor for measuring pH it has important characteristics like its robustness and durability, small size (contact area < 1 mm^2), massive production potential, low output impedance, and low response time (1000 times minor in comparison with the glass electrode) [1, 6].

The ISFETs used in this work have been fabricated at the Instituto de Microelectrónica de Barcelona (IMB-CNM) clean room by using standard microfabrication techniques. It is a Si_3N_4 gate NMOS ISFET chip with 3x3 mm size. The chip is fixed in a printed circuit board and wire bonded to the pads for electrical connection. A polymer film is deposited by photolithographic techniques to cover all the electrical parts leaving open the gate area for liquid contact. To close the circuit an Ag/AgCl reference double junction electrode is used.

2.2. System's design

The system is basically composed by two stages: the signal conditioning stage and the stage for processing, visualization and transmission of the data acquired by the sensor.

The conditioning stage should guarantee the adequate polarization of the microsensor. During the design the polarization circuit was carefully selected to avoid diverse unwanted effect. By this reason the polarization of the microsensors was done in the feedback mode, in which the compensation of the gate-source potential variation (V_{gs}) is done by a feedback signal, applied through the reference electrode [5]. The polarization must guarantee a constant drain-source current (I_{ds}) and a constant drain-source voltage (V_{ds}) as well. As usual, the values used for I_{ds} and V_{ds} were 100 μA and 500 mV, respectively.

Figure 2 shows the electrical circuit used in the design of the system's conditioning stage.

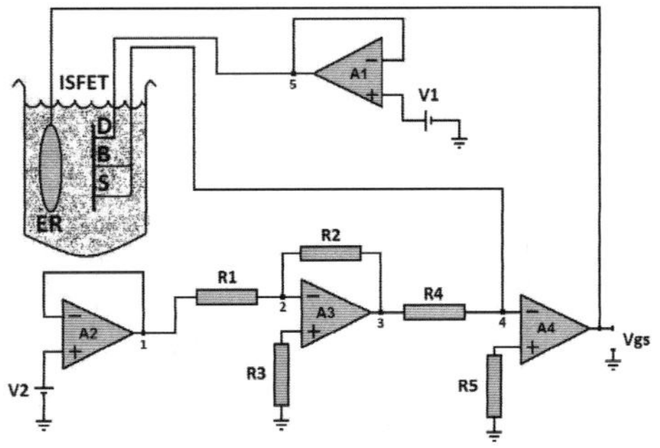

Figure 2. Electrical circuit used in the design of the system's conditioning stage.

The microcontroller CY8C29466 [8] was used as the basic processing unit. This microcontroller belongs to the family of PSoC, devices of small size and low cost. The most remarkable property of this device is its capacity to integrate analog components in the digital section of the chip [9]. This characteristic allowed the partial implementation of the polarization circuit on the PSoC, making possible the integration and portability of the system. Other interesting characteristics of the PSoC microcontroller are: 24 MHz Central Processing Unit (CPU) M8C, Harvard 8 Bit architecture, 32 K of FLASH memory and 2 K of SRAM, 12 blocks in the analog section and 16 blocks in the digital one [8, 9]. The schematic design of the developed system is shown in figure 3.

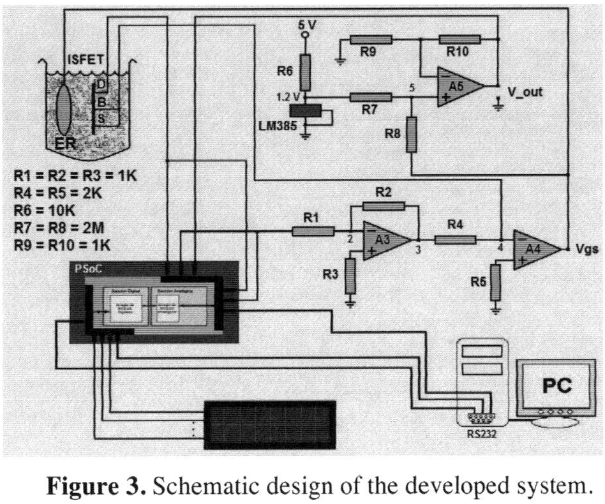

Figure 3. Schematic design of the developed system.

To reach a wide nominal range for V_{gs} measurement, it is important that the system can read negative values. For this reason it was necessary to modify the PSoC input signal as this device cannot measure negative voltages. To overcome this inconvenience, the amplifier A5 was included. This amplifier must be configured as a non-inverter summing to get the A5 output voltage equal to $Vgs+V_{LM385}$, with V_{LM385}, being equal to 1.2 V. In this way, even if the input nominal range for the

PSoC is always from 0 to 2.6 V [9], the nominal range of the system will be from -1.2 to +1.4 V, which allows measuring negative values of Vgs. In the diagram of figure 3 the operational amplifiers A3 and A4 are not part of the polarization circuit implemented in the PSoC.

The visualization of the data is carried out by means of a liquid crystal display (LCD) interconnected with the microcontroller. The FET microsensors obtained data is stored in the PSoC by the module E2PROM implemented in it. The transmission of these data towards a PC is made by a RS -232 serial port. The system reads the values of V_{GS}, V_{DS} and I_{DS} with the PSoC. The values of V_{DS} e I_{DS} are only acquired to check the stability of the sensor operation point. The value of V_{GS} is given as a function of the analyzed sample pH.

The configuration of the main microcontroller parameters and the integrated components was done using the PSoC Designer v4.3 software. For the programming, C language was used due to the advantages this language offers, among others, the facility to work with floating numbers and the option for re-programming the code or upgrading it. The modules implemented in the PSoC are shown in the figure 4. These modules are referred to a pair of digital/analogical converters (DAC); the E2PROM module to emulate this data storage memory in the microcontroller FLASH memory; a LCD module that establishes the interface with the LCD; three programmable gain amplifiers (PGA); an three entrances and three exits analog/digital converter (TRIADC), and the module UART for the communication with the serial port.

Figure 4. Modules implemented in the PSoC.

3. Results

A group of measurements for the calibration of the electronic designed system, was made after its implementation in a physical support system.

With the aim of verifying the main parameters of the polarization circuit, a simulation with the help of the design software Proteus v7.10, was carried out. By means of this simulation, it was possible to verify that the designed configuration gave the correct values of the operation point of the sensor, and it was also possible to verify that the V_{gs} values were under the required range. The measurements sweep a tension range between -1V and +1V in each canal. These measurements allow knowing systematic errors like sensitivity and zero errors in order to minimize them by software manipulations.

Figure 5 shows the comparison between the voltage values measured and displayed in the LCD of the system after calibration (V_PSoC), versus the values measured with a standard voltage measuring instrument (V_Standard). The data processing and the graphic shown in the figure 5 were carried out with STATGRAPHICS v5.1 software. The results obtained by the data processing software are shown in table 1.

The linear fitting, with linear correlation coefficient of 0.999997, is indicative of the excellent relation between the values shown in the LCD of the system, and the real values measured. It means that after the calibration the gain errors have practically null values. Furthermore, from the intercept value we can conclude that the errors related to zero have values of less than 2 mV.

Table 1. Results offered by the data processing software.

Calibration Parameters	Results
Slope (m)	1.0009 ± 0.0002
Intercept (n)	-1.9676 ± 0.1149 mV
Linear correlation coefficient (R^2)	0.999997

All these results demonstrate that the measurement system developed with the PSoC has the appropriate accuracy.

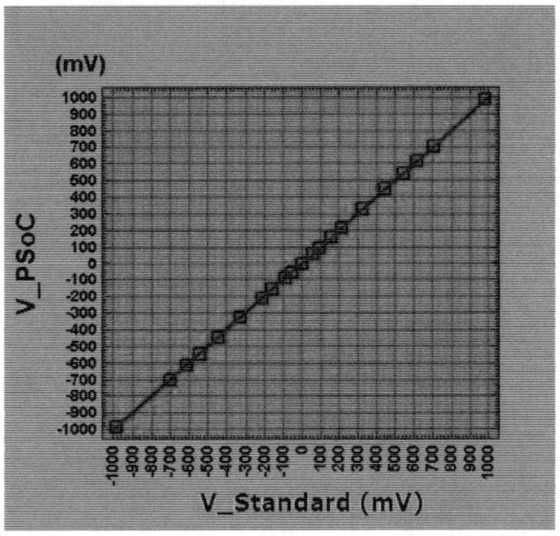

Figure 5. Measurement of V_{GS} after the calibration.

4. Conclusions

A portable measurement system for FETs microsensors based on a PSoC CY8C29466 microcontroller as the main element of the hardware was developed and validated. The use of the PSoC has contributed to considerably reducing the cost, size, and weight of the system while maintaining good performance, autonomy and flexibility. The portable system has an excellent linearity, accuracy and is also able to storage data for further transfer to a PC serial port using the PSoC facilities. The nominal range of voltage to be measure by this portable measurement system include negative and positive values from -1.2 V to +1.4 V.

Acknowledgments

Authors wish to acknowledge the CYTED Project 510AC0408 "REDSENS" for the funding of collaboration stages.

5. References

[1] Jiménez C, Domínguez C, Arias-de Fuentes O, Lastres A and Valdés E 2006 Microsensores de estado sólido tipo ISFET: Estructura y fundamentos *Sensores y Microsistemas* **2**, (Buenos Aires: Garbarino Hnos)
[2] Jiménez C, Orozco J and Baldi A 2010 *Sensors* **10** 61-83
[3] Hodge AM and Newcomb RW 2002 Evaluation of the VLSI Adaptation of the Chemfet, a Biosensor for Fluid Analysis *Proc. Circuit and Systems ISCAS 2002 (IEEE International Simposium*, Vol II) pp 580-583
[4] Bergveld P 2003 *Sensor Actuat. B* **88** 1-20
[5] Blanco, A 2009 *Sistema con PSoC para la caracterización de microsensores basados en FET.* MSc.Thesis (Havana: University of Havana)
[6] Casans S *et al* 2001 *Sensor Actuat. B* **76** 629-633
[7] Orozco J *et al* 2007 *Meas. Sci. Technol.* **18** 935-940
[8] Cypress Semiconductor, 2006 *PSoC Mixed signal array* CY8C29466 - CY8C29666 *Datasheet*,
[9] Ashby R 2005 *Designer's Guide to the Cypress PSoC™* (New York: Elsevier)

Determination of heavy metals contamination using a silicon sensor with extended responsive to the UV

M Aceves-Mijares[1], J M Ramírez, J Pedraza, S Román-López and C Chávez
Instituto Nacional de Astrofísica, Óptica y Electrónica, Apdo. 51, Puebla, Pue. México 72760

E-mail: maceves@inaoep.mx, jmram@inaoep.mx, jpch@inaoep.mx

Abstract. Due to its potential risk to human health and ecology, the presence of heavy metals in water demands of techniques to determine them in a simple and economical way. Currently, new developments of light emitters and detectors open a window of opportunities to use optical properties to analyze contaminated water. In this paper, a silicon sensor developed to extend its sensitivity up to the UV range is used to determine heavy metals in water. Cadmium, Zinc, Lead, Copper and Manganese mixed in pure water at different concentrations were used as test samples. The photocurrent obtained by the light that passes through the samples was used to determine the optical transmittance of pure and contaminated water. Preliminary results show a good separability between samples, which can be used for qualitative and quantitative detection of such heavy metals in water.

1. Introduction

The presence of heavy metals in water, natural or due to environmental pollution, has received considerable attention of the research community in the last years due to its potential risk to human health and ecology. The toxicity of heavy metals such as cadmium, zinc, lead, mercury, copper, manganese, and others, can have harmful and even lethal effects on the human body, depending on the concentration and frequency of consumption. Regarding environmental aspects, recent studies indicate an increasing of heavy metals concentrations in lakes or rivers, frequently correlated with an increased industrial activity in the region [1, 2, 3]. As a result, techniques and methods for monitoring concentration of toxic metals within legally defined limits acquire special relevance.

A number of techniques have been developed over the years for analysis of heavy metals concentrations in water, many of them based on spectrophotometric methods. Among these, atomic absorption spectrometry (AAS), inductively coupled plasma (ICP)-atomic emission spectrometry and ICP-mass spectrometry have been broadly reported with several variants. AAS assesses concentration of analyte in a sample based on the absorption of optical radiation by free atoms. The radiation flux with and without a sample is measured, the absorbance is then translated to analyte concentration using the Beer-Lambert Law [4]. In atomic emission spectrometry (AES) the analyte is exposed to a flame, and the thermal energy excites the atoms into electronic states that emit light when they return to the original state. Radiation wavelength and intensity provide information about identity and amount of the element in a sample [5]. ICP-AES makes use of inductively coupled plasma to produce excited atoms, which emit the radiation detected in the spectrometer. In reference [6], flame atomic absorption spectrometry (FAAS) is used for the determination of some heavy metals, presenting a new

[1] To whom any correspondence should be addressed.

Published under licence by IOP Publishing Ltd

cloud point extraction procedure for the separation and preconcentration of metals in water and food samples. In reference [7] the authors present a novel continuous powder introduction microwave-induced plasma atomic emission spectrometry method (CPI-MIP-AES), for trace determination of metals in ground and tap water samples after preconcentration on activated carbon. Another technique for the analysis of heavy metals in liquids is Laser Induced Breakdown Spectroscopy (LIBS). In this technique a highly energetic laser pulse is focused onto the sample producing a microplasma with molecular and atomic excited states, which will emit the expected radiation to be detected in the spectrometer. Reference [8] presents a method using LIBS in which a wood slice is used as a liquid absorber to transform liquid sample analysis to solid sample analysis. Another liquid to solid sample analysis using LIBS is presented in reference [9], where liquid solutions are converted into solid pellets of calcium hydroxide by mixing with calcium oxide. X-ray fluorescence spectrometry (XRFS) is another well-known technique, which has been used to analyze heavy metals content in water. XRFS is based on the emission of fluorescent X-rays from a sample, which has been excited by bombarding with high-energy X-rays. References [10] and [11] present studies using XRFS. Detection of heavy metals in water using surface plasmon resonance spectroscopy (SPRS) combined with anodic stripping voltammetry is presented in reference [12]. SPRS is based on the principle of resonance when the frequency of incident light matches the natural frequency of surface electrons oscillating against the restoring force of positive nuclei. Another technique often used in water analysis is Ion Chromatography (IC). Experiments on detection of heavy metals in water using IC with ultraviolet detection are reported in references [13, 14].

Besides all the efforts already mentioned, there is a tendency to take advantage of new light emitters and sensors to provide simplest and economical analytical methods. For example, in reference [15] our group has reported a silicon sensor that extends the silicon response, normally in the visible to near infrared (VIS-NIR), to the ultraviolet (UV) region. The sensor combines a PN junction with a film of Silicon Rich Oxide (SRO) with the property of converting UV radiation into VIS-NIR light.
SRO obtained by Low Pressure Chemical Vapor Deposition (LPCVD) has demonstrated to have a higher emission of visible to near infrared emission when illuminated with UV light [16]. Figure 1 shows the wavelength range used to obtain the SRO emission, also in the inset is presented the typical photoresponse of a silicon diode and the SRO emission. As can be observed, the wavelength coincidence between the emission and the detection can be used to extent the silicon range of detection. So, a silicon sensor has been developed that prolongs the response of the silicon up to 200 nm.

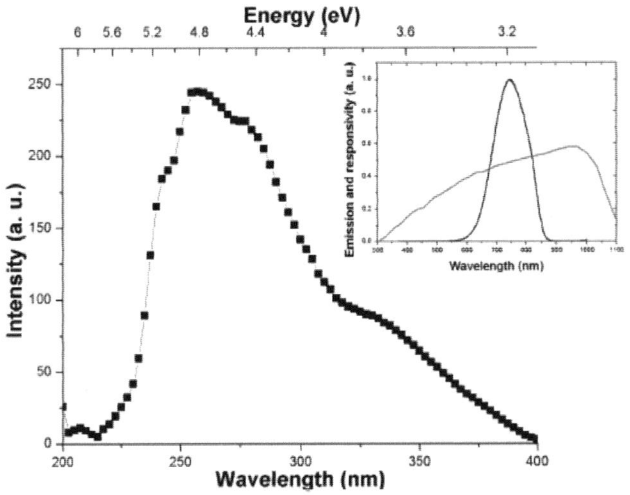

Figure 1. Wavelength range used to stimulate a SRO film producing an emission centered at 750 nm. The inset shows the characteristic SRO emission and the typical photoresponse of a silicon PN diode (red line).

In this work, a preliminary study aiming to detection of heavy metals in water based on analysis of transmittance of incident ultraviolet light is presented. For that purpose, an experimental arrangement for measurement of transmitted light in the 200 to 400 nm UV range was developed. Cadmium, Zinc, Lead, Copper and Manganese in pure water solution at different concentrations were used as test samples. Preliminary results show a good separability between samples, which can be used for qualitative and quantitative detection of such heavy metals in water. Also, fabrication details and characteristics of the Si sensor with responsivity from 200 to 1000 nm are shown.

2. Experimental

The Si sensor was built on (100) P type epitaxial silicon wafers, the anode was done either by implanting or by thermal diffusion of phosphorous into the P type wafer. On the PN junction a SRO layer was deposited [17]. The active area used in this study was 1500 square micrometers. Average optical power was measured using an International Light model IL1400 radiometer. The currents were measured with a Keithley model 6517A electrometer. The set up used to measure the relative transmittances is shown in figure 2. For more detail on the experimental procedure refer to [18]. The relative transmittance, Tr, was measured in the wavelength range of 200 to 900 nm. However, data in the range of 200 to 400 nm is enough to determine the contamination. The relative transmittance (Tr) was defined as the ratio of the photocurrent produced by light passing through contaminated water (Ic) and that of pure water (Ip) as shown in equation (1). An example of relative transmittance is shown in figure 3.

$$T_r = \frac{I_C}{I_P} \tag{1}$$

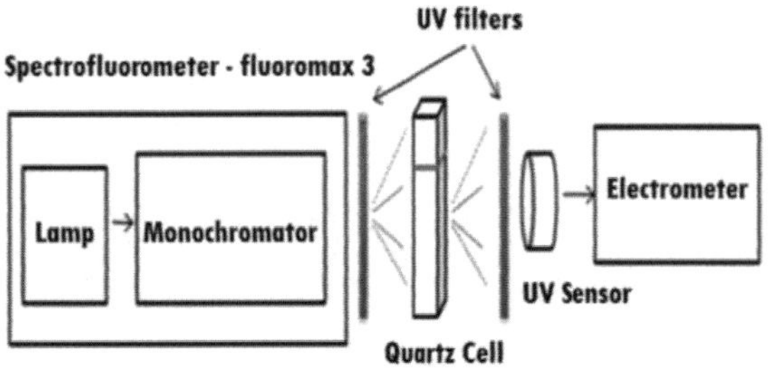

Figure 2. Experimental setup, the Xe lamp and the monochromator of the spectrofluorometer is used as a monochromatic light source.

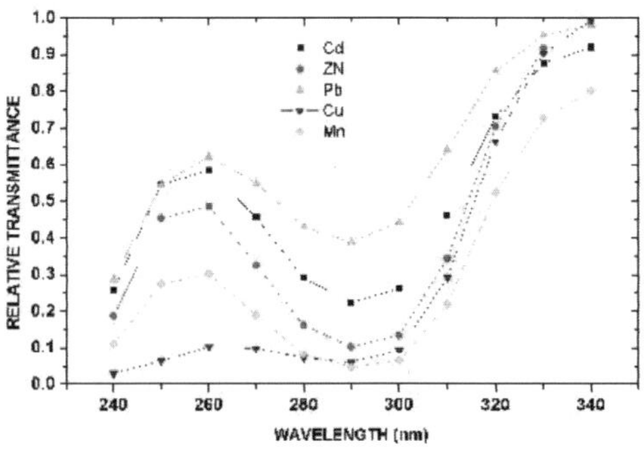

Figure 3. Relative transmittance as function of incident wavelength for the five metals used with 5000 ppm.

3. The sensor

Figure 4a shows the current density as a function of the wavelength for our sensor and two commercial Si sensors. The dark current density in our devices is in the order of pA/cm^2. The well-known relationship in equation (2) was used to determine the responsivity of our UV silicon sensor, UVSis. In this formula Jp is the photocurrent density and Po is the optical power density.

$$R = \frac{J_P}{P_0} \tag{2}$$

Figure 4b shows the typical responsivity measured for the UVSis. As can be seen, from 200 nm the response increases having a maximum around 300 nm. Then, the responsivity starts to reduce in agreement with figure 1. After 400 nm a PN silicon photodetector standard response is observed, therefore demonstrating that the SRO layer really contributes to the sensor sensitivity in the UV region.

Figure 4b can be used to estimate the efficiency of emission of SRO. The sensor's responsivity in the range of 700 nm is 0.3 A/W, meaning that per each watt of light impinging in the sensor, a 0.3 A photocurrent will be obtained. Now, if 1 W of 270 nm radiation shines on the sensor that will be converted into 700 nm light that produces only 0.1 A. It has to be considered that not all light converted from UV into VIS will reach the sensor's surface. Let us suppose that only 50% of the generated VIS light will reach the sensor, then a photocurrent of 0.15 A should be obtained. However, only around 70 %, 0.1 A are obtained. That is an indication that the SRO film is a really good wavelength shifter, converting most of the UV radiation into visible light.

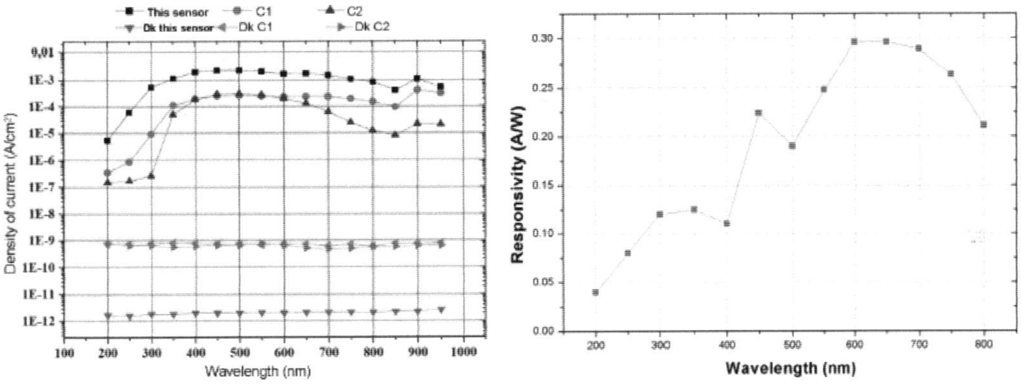

a) b)

Figure 4. a) Current density and b) responsivity as a function of wavelength for the UVSis. The current density is compared with two commercial Si diodes.

4. Heavy metal detection

The relative transmittance, $T_{r,}$ was measured in the wavelength range of 200 to 400 nm. The results show clearly discernible curves for the different metals and concentrations. Figure 3 shows the transmittance curves for the different metals used with a concentration of 5 000 ppm. 100 ppm was the minimum quantity clearly determined. In this wavelength range, the T_r and the concentration follows a quadratic relationship. In a simple way this relationship can be used to determine the metal and its concentration, experimentally an error of less than 5% was found.

In order to automatically classify the different contaminated solutions, a database was created with the relative transmittance obtained at five different wavelengths in the range 250-300 nm, which were found to give the most discriminative values associated to the greatest variances. Separability on data clustering was tested through Euclidean distance classification, which is a well-known pattern recognition technique. The object under test is assigned to the nearest prototype among the p classes using the minimum-Euclidean-distance criteria, equation 3, where each prototype is obtained as the mean vector of the m samples in each class, equation 4. In this work, the number of classes corresponds to p = 5 (five contaminating metals), and m = 3 (number of samples associated to each contaminating metal). Figure 5 shows the Euclidean distance obtained from the samples in the database to each prototype. Inter and intra-class statistical analysis indicates that an adequate data clustering with good data separability can be obtained through the described method. Further experimentation using non-linear classification based on neural networks is currently in progress.

$$d_{ip} = \sqrt{\sum(\bar{X}_p - \vec{x}_i)^2} \tag{3}$$

$$\overline{X}_p = \frac{1}{m}\sum_{j=1}^{m}\vec{x}_j \tag{4}$$

Additional experimentation aiming to the generation of a comprehensive database should be carried out in order to come up with a complete technique evaluation. Those experiments should include more contaminants at different concentrations, as well as combinations of them, which could take advantage of the non-linear classification capabilities of the neural network.

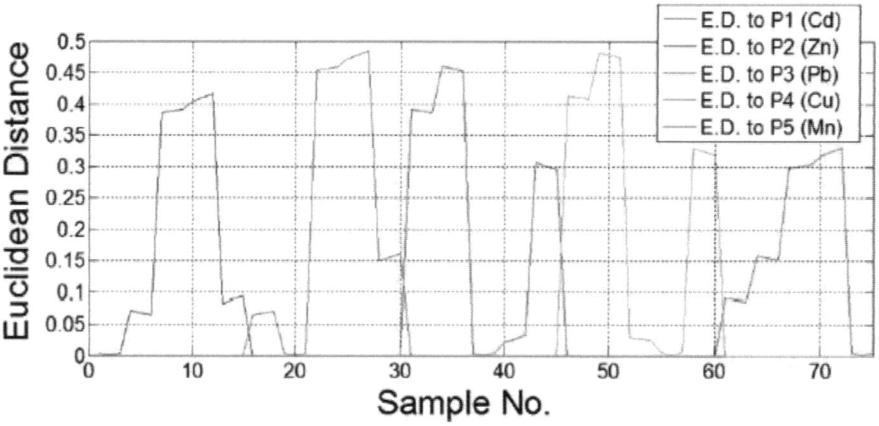

Figure 5. Euclidean distance from database samples to the five prototypes.

5. Conclusions

A silicon sensor with responsivity from 200 to 1000 nm was used to determine the transmittance of pure and contaminated water. The sensor combines the light emission properties of a Silicon Rich Oxide film with the detection capabilities of a PN silicon junction. The efficiency of emission was estimated from the measured responsivity. The optical transmittance of Cadmium, Zinc, Lead, Copper and Manganese mixed in pure water at different concentrations were measured from 200 to 900 nm. However, only data obtained from 240 to 340 nm was enough to obtain relevant information to determine the contaminant. Separability on data clustering was tested through Euclidean distance classification in order to automatically classify the different contaminated solutions. Our preliminaries results, using pure water and one contaminant at a time, show that it is possible to determine quantitatively and qualitatively heavy metals such as Cadmium, Zinc, Lead, Copper and Manganese in concentrations up to 100 ppm.

Acknowledgements. The authors appreciate the support of CONACyT.

6. References

[1] Diagomanolin V, M. Farhang V M, Ghazi-Khansari M and Jafarzadeh N 2004 *Toxicology Letters* **151** (1) 63-67
[2] Kishe M A and Machiwa J F 2003 *Environment International* **28** (7) 619-625
[3] Rios-Arana J V, Walsh E J and Gardea-Torresdey J L 2004 *Environment International* **29** (7) 957-971
[4] Welz B and Sperling M 2007 *Atomic Absorption Spectrometry* (Third Edition, Wiley-VCH Verlag GmbH, Weinheim, Germany)
[5] Lajunen L H J and Peramaki P 2004 (2nd Ed., Pub. Royal Society of Chemistry, Thomas Graham House, Science Park, Milton Road, Cambridge, UK)
[6] Citak D and Tuzen, M 2010 *Food and Chemical Toxicology* 48 (5) 1399-1404

[7] Jankowski K, Yao J, Kasiura K, Jackowska A and Sieradzka A 2005 *Spectrochimica Acta Part B: Atomic Spectroscopy* **60** (3) 369-375

[8] Chen Z, Li H, Liu M and Li R 2008 *Spectrochimica Acta Part B: Atomic Spectroscopy* **63** (1) 64-68

[9] Daz Pace D M, D'Angelo C A, Bertuccelli D and Bertuccelli G 2006 *Spectrochimica Acta Part B: Atomic Spectroscopy* **61** (8) 929-933

[10] Capitelli F, Colao F, Provenzano M R, Fantoni R, Brunetti G and Senesi N 2002 *Geoderma* **106** (1-2) 45-62

[11] Zarazua G, Avila-Perez P, Tejeda S, Barcelo-Quintal I and Martinez T 2006 *Spectrochimica Acta Part B: Atomic Spectroscopy* **61** (10-11) 1180-1184

[12] Wang S, Forzani E S and Tao N 2007 *Anal. Chem.* **79** (12) 4427-4432

[13] Michalski R 2004 *Polish Journal of Environmental Studies* **13**(1) 73-77

[14] Santoyo E, Santoyo-Gutierrez S and Verma S P 2000 *Journal of Chromatography A* **884** (1-2) 229-241

[15] Berman-Mendoza D, Aceves-Mijares M, Berriel Valdos L R, Carranza J, Pedraza J, and Dominguez-Horna C, Falcony C 2005 *Laser Focus World* 103-105

[16] Berman D, Aceves, Gallegos A, Morales A, Berriel L R, Carrillo J, Flores F, Falcony C, Dominguez C, Llobera A, Riera M and Pedraza J 2004 *phys. stat. sol. (c)* **1(S1)** S83-S87 DOI 10.1002/pssc.200304864

[17] Aceves-Mijares M, Berman-Mendoza D, Berrier Valdos R, Dominguez-Horna, Yu Z and Falcony C2006 *IBERSENSOR Revista Iberoamericana de Sensores* **5** (1) http://www.inaoep.mx/revista-electronica/index.php

[18] Chavez-Vallarino C, Mariano Aceves-Mijares M and Pedraza-Chavez J 2011 *8th International Conference on Electrical Engineering, Computing Science and Automatic Control* (Merida, Yucatan, Mexico) pp. 1066-1069

Seasonal effects in arsenic levels in drinking water in the Lagunera region

Aldo Uriel Aguilar-Muñiz[1,3], **Francisco Valdés-Perezgasga**[1] and **Gonzalo G García-Vargas**[2]
[1]Instituto Tecnológico de la Laguna, 27000, Torreón, Coahuila, México
[2]Facultad de Medicina, Universidad Juárez del Estado de Durango, 35050, Gómez Palacio, Durango, México

E-mail: doal86@hotmail.com

Abstract. Changes in arsenic content in the drinking water in the Lagunera Region in northern Mexico are presented here. Arsenic and water conductivity were measured both before and after the yearly irrigation cycle and also before and after a river bed that is usually dry, received water in a 2010, a wet year. The results were processed with a Geographic Information System (GIS) to understand better the changes in relation to the dry riverbed. The present work can evolve towards a real-time, sensor-based system.

Keywords: Arsenic, Hydroarsenicism, Geographic Information Systems.

1. Introduction

Arsenic (As) is a toxic element. It may cause serious health problems even when ingested in low quantities for a prolonged time. Hence the relevance of monitoring its levels in drinking water. Its health effects go from simple skin alterations to cancer of the bladder, skin, kidneys, lungs, liver and prostate [1]. When arsenic enters the organism through the digestive system it stays in the blood for a short time as it spreads to the liver, kidneys and digestive tract. A fraction of the arsenic is excreted in urine both in its inorganic and its methylated form. However, some of it stays in the body and may accumulate overwhelming the detoxification mechanisms of the body [2].

2. Hydroarsenicism

There are several places on the planet where hydroarcenicism is present. It arises due to a particular geology or because of human activities such as mining and other industries activities and also as a response to aquifer depletion. The Lagunera Region in north-central Mexico, in the states of Durango and Coahuila, is one of these regions.

Recent research has found that the origin of the arsenic present in groundwater in the Lagunera Region responds to aquifer depletion. This in turn is caused by the damming and diversion of its rivers and a concurrent overdraft of groundwater [3]. The Lagunera Region is one of the driest in Mexico (annual rainfall is 260mm) [4]. Being in the center of the vast Chihuahuan Desert, it has scarce water resources that include the Nazas and Aguanaval rivers, both of them Endorrheic Rivers that used to form vast seasonal lagoons that no longer exist. Most of the extracted groundwater is destined to grow

[3] To whom any correspondence should be addressed.

alfalfa and other forage crops. This responds to the Lagunera Region being the main milk-producing region in Mexico.

The presence of arsenic in groundwater in the Lagunera Region has been recognized since the 1970s [5]. In 2007, López-Uriostegui found arsenic levels in drinking water that far exceeded the Mexican norm of 25 μg/L. The highest value in a well was found in the Holanda Ranch, 10 km south of the city of Tlahualilo, Durango with an arsenic content of 643 μg/L [6].

Irrigation agriculture is dominant in the Lagunera Region. During the irrigation season, water is released from the dams and it flows through the normally dry riverbeds. While moving through the riverbed some water infiltrates to the aquifer. Later, water is moved through cement-paved canals to the fields. Irrigation itself allows for further infiltration of water from the fields to the aquifer. It is expected that during this time "young" water with low arsenic levels may have an effect on arsenic levels in the groundwater. Furthermore, the rain season occurs between May and October, coinciding with the irrigation cycle. Rain may also have a similar effect on arsenic levels in the aquifer.

3. Methods

Fifty samples of drinking water were taken in different points of the metropolitan area of the Lagunera Region (in the cities of Gómez Palacio, Lerdo and Torreón). The sampling dates were selected to bracket both the rainy and the irrigation season. According to data supplied by the National Water Commission, the rainy season occurs between May and October as in the rest of the Mexican highlands. During these dates the dams release water for irrigation in well-defined periods. Samples were taken in April and then again in October. For every sample, latitude and longitude data were registered with a GPS-III personal apparatus (Garmin International Inc., Olathe, Kansas).

The aquifer is expected to receive water both through the riverbed before the water reaches the canals and then through the irrigated fields. Showers that may occur locally during this time also contribute to the recharge of the aquifer.

4. Laboratory analyses

Arsenic levels were analyzed in the Toxicology lab at the Medical School of the Durango Juarez State University using atomic absorption spectrometry with a coupled hydride generator. As a complement, conductivity was measured in each of the samples as a way of seeing the effect of overall salt content. Conductivity was measured using a NEO-2 portable precision conductivity meter (Mesa Laboratories, Lakewood, Colorado) capable of measuring conductivities from 0.001 mS/cm to 1 μS/cm.

After arsenic and conductivity measurements were done, the results were analyzed using the ArcGIS geographic information system (ESRI, Redlands, California). This analysis allowed the visualization of the probable distribution of arsenic in groundwater.

5. Results

Images of the probable distribution of arsenic in groundwater in the metropolitan area of the Lagunera Region were produced. These images were made using sample information before and after the irrigation cycle and the rainy season. This study was done only for the urban area. A satellite photograph of the urban area was used as the base layer of the image as seen in figures 1-2. A second layer included the sampling points. A third layer was a raster that results from the spatial interpolation of the arsenic levels found in each one of the fifty samples. Red areas correspond to levels higher than the Mexican norm of 25 μg/L [7]. It is worth noting that the World Health Organization recommends an arsenic content of no more than 10 μg/L [8].

Figure 1 shows the resulting image built from the samples of April 2010 prior to the irrigation cycle. For this period the highest value was 43.325 μg/L of arsenic in sample 8, the lowest value was 8.625 μg/L of arsenic in sample 34. The mean of all 50 samples was 23.25 μg/L of arsenic.

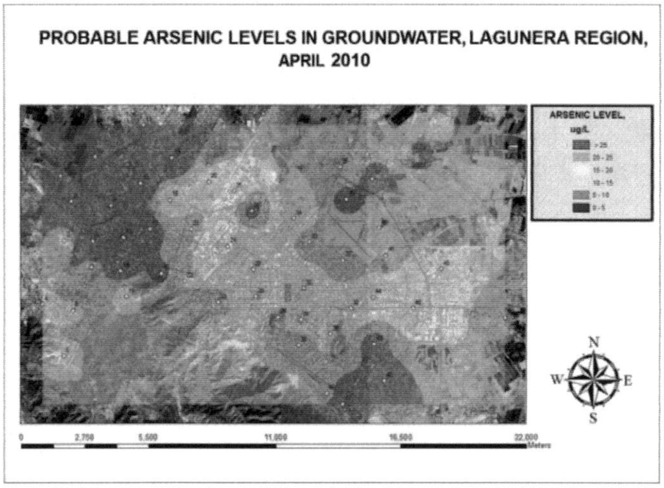

Figure 1. Probable arsenic levels of groundwater in the metropolitan area of the Lagunera Region in April, 2010.

Figure 2 shows the image that results from the samples taken in October 2010, after the irrigation cycle. For this period the highest value was 70.960 μg/L of arsenic in sample 25, the lowest value was 11.075 μg/L of arsenic in sample 49. The mean of all 50 samples was 26.65 μg/L of arsenic.

Figure 2. Probable arsenic levels of groundwater in the metropolitan area of the Lagunera Region in October, 2010

The results of the conductivity studies are shown in Figures 3 and 4. Although there is not a mention of conductivity in the Mexican norms regarding drinking water, we found that the zones with the highest conductivity coincide with the zones with the highest arsenic levels. This can be a cross-check of sorts as conductivity stands for overall salt content including arsenic salts. This zone is in the northeastern part of the image and is located in the city of Gómez Palacio in the state of Durango.

In figures 3 and 4 the difference between the resulting April and October images do not show a stark contrast as is the case with the arsenic images. This may point to the fact that arsenic salts are not the dominant component of conductivity in the groundwater of the Lagunera Region. And therefore, that there is not a seasonal effect in conductivity.

Figure 3. Probable conductivity of groundwater in the metropolitan area of the Lagunera Region in April, 2010

Figure 4. Probable conductivity of groundwater in the metropolitan area of the Lagunera Region in October, 2010

6. Conclusions

From figures 1 and 2 it can be noted that the irrigation cycle had an effect in diminishing arsenic levels in the eastern part or the metropolitan area but increasing it in the western part. It is interesting to note that the dry bed of the Nazas River coincides with the eastern limit of the red zone in figure 2, this the zone where arsenic above 25 μg/L is present. In 2010 there was rainfall above the average that forced the authorities to release water in a normally dry portion of the Nazas river in addition to the water moved through the canals. Since 1968 when the Francisco Zarco dam was completed, this has only happened in 1992 and 2008. The rise in arsenic levels is counterintuitive as "new" water brought by rain and the river (through the normal irrigation cycle and the extraordinary 2010 flood) is expected to dilute arsenic levels in the aquifer.

The 2010 flood and the ensuing passage of the water through the normally dry riverbed may have provoked a violent local recharge of the aquifer and a corresponding alteration of the phreatic layers that may have caused the arsenic to rise. This effect may have manifested with special force near the riverbed but not in zones farther from it, i.e. the eastern portion of the metropolitan area. Another explanation is that the floodwaters may have been laced with arsenic. There are a number of mines (and their tailings dams) in several tributaries of the Nazas upstream from Torreon.

7. Recommendations

Arsenic in drinking water is a public health issue. Furthermore, it is problem that has been worsening as time passes as the main cause of it, the overdrafting of groundwater is accelerating. It is therefore necessary that this problem be addressed even when it will take time to revert to an earlier state when arsenic in water was not an issue as drafting was kept below recharge.

More analyses are needed both from events in the future but also from data from previous years. This would help build a more complete picture of the recharge dynamics of the local aquifer. However, the present work justifies further studies towards the development of sensor-based, automatic monitoring systems that can evaluate arsenic levels both in the aquifer and the water for irrigation coming from the dams on a daily basis. This would help detect illegal toxic spills from mines upstream, take aquifer management decisions more speedily, and communicate risk to the 1.5 million inhabitants of the metropolitan area of the Lagunera Region regarding the safety of tap water.

8. References

[1] Cullen W.R. and Capper P 2008 Is arsenic an aphrodisiac? The sociochemistry of an element. *Royal Society of Chemistry,* London

[2] Ortiz Cruz V. Comparación de dos métodos para la determinación de arsénico en agua de los pozos de la costa agrícola de Caborca, Sonora. Instituto Tecnológico de Sonora, Thesis number 19631. http://tesis.uson.mx/digital/tesis/wsp_tesis_detalle_nuevo.asp?tesis=19631 [Consulted on June 4th, 2011]

[3] Ortega-Guerrero M A 2004 Origin of high concentrations of arsenic in groundwater at the "La Laguna Region", northern Mexico, and implications on aquifer management. *Proc 32nd Int Geological Congress*, Florence, Italy, August 20-28, 2004, pp. 1486

[4] Comisión Nacional del Agua. Precipitación media anual 1941-2005 http://smn.cna.gob.mx/index.php?option=com_content&view=article&id=12&Itemid=77 [Consulted on August 26th, 2011]

[5] Albores A, Cebrián M E, Téllez I and Valdez B 1979 Estudio comparativo de hidroarsenicismo crónico en dos comunidades rurales de la Región Lagunera de México. *Boletín de la Oficina Panamericana de la Salud* **86** 196

[6] López Uriostegui Z 2007 *Determinación de arsénico en aguas subterráneas de la Comarca Lagunera*. B.Sc. Thesis. Instituto Tecnológico de Mazatlán. Mazatlán, Sinaloa, México

[7] Secretaría de Salud 2000 Modificación a la Norma Oficial Mexicana NOM-127-SSA1-1994, *Salud ambiental. Agua para uso y consumo humano. Límites permisibles de calidad y tratamientos a que debe someterse el agua para su potabilización*. Diario Oficial de la Federación, October 20, 2000. http://www.salud.gob.mx/unidades/cdi/nom/m127ssa14.html [Consulted on 30 de Junio del 2011]

[8] World Health Organization. *Guías para la calidad del agua potable*.Vol.1: Recomendaciones. Tercera edición. ISBN: 92 4 154696 4 [Consulted on August 23, 2011] http://www.who.int/water_sanitation_health/dwq/gdwq3_es_full_lowsres.pdf

Measurement system for an in-vitro characterization of the biomechanics and hemodynamics of arterial bifurcations

D Suárez-Bagnasco[1,4], G Balay[1], L Cymberknop[3], RL Armentano[2] and CA Negreira[1]

[1]Laboratorio de Acústica Ultrasonora, Instituto de Física, Facultad de Ciencias (FCIEN), UDELAR, Montevideo, Uruguay.
[2]Departamento de Fisiología, Facultad de Medicina (FMED), UDELAR, Montevideo, Uruguay.
[3] Departamento de Electrónica, Facultad Regional Buenos Aires, Universidad Tecnológica Nacional (UTN), Buenos Aires, Argentina.

E-mail: dsb@ieee.org

Abstract. Arterial behaviour in-vivo is influenced, amongst other factors, by the interaction between blood flow and the arterial wall endothelium, and the biomechanical properties of the arterial wall. This interaction plays an important role in pathogenic mechanisms of cardiovascular diseases such as atherosclerosis and arteriosclerosis. To quantify these interactions both from biomechanical and hemodynamical standpoints, a complete characterization and modelling of the arterial wall, blood flow, shear wall and circumferential wall stresses are needed. The development of a new multi-parameter measurement system (distances, pressures, flows, velocity profiles, temperature, viscosity) for an in-vitro characterization of the biomechanics and hemodynamics in arterial bifurcations (specially in carotid bifurcations) is described. This set-up represents an improvement relative to previous set-ups developed by the group FCIEN-FMED and is presently under development. Main subsystems interactions and environment-system interactions were identified and compensated to improve system's performance. Several interesting problems related with signal acquisition using a variety of sensors and some experimental results are shown and briefly discussed. Experimental data allow construction of meshes and parameter estimation of the biomechanical properties of the arterial wall, as well as boundary conditions, all suitable to be employed in CFD and FSI numerical simulation.

1. Introduction

Arterial behavior in-vivo is influenced, amongst other factors, by the interaction between blood flow and the arterial wall endothelium, and the biomechanical properties of the arterial wall.
This interaction plays an important role in pathogenic mechanisms of cardiovascular diseases such as atherosclerosis and arteriosclerosis. To quantify these interactions both from biomechanical and hemodynamical standpoints, a complete characterization and modeling of the arterial wall, blood flow, shear wall and circumferential wall stresses are needed. Characterization can be done in-vivo and in-

[4] To whom any correspondence should be addressed.

Published under licence by IOP Publishing Ltd

vitro. In-vivo, by means of invasive and non-invasive (tonometry, pulsed Doppler and pulse wave velocity) methods. In invasive methods, suitable instrumented specimens are used to perform the measurements on a given arterial segment. The specimen is selected depending of the objectives of the experimental work. In-vitro, the characterization can be done by means of a hemodynamic workbench simulator (HWBS) in which arterial segments are intercalated in closed fluid circuits that emulates the main features of systemic arterial circulation [1-3]. A sound in-vitro characterization of arterial segments (including pressure and flow patterns) may furnish valuable information to be applied in-vivo [4]. A specific experimental set-up is needed in order to be able to work with arterial bifurcations. Measurements obtained from cryopreserved [5] and fresh arterial bifurcation samples (for example carotid bifurcation amongst others) subjected to different hemodynamic regimes allow the characterization of the biomechanical properties of arterial segments. These are obtained by parameter estimation and will be employed in a model for the numerical simulation of the interaction between structure and fluid. Velocity profiles obtained in the workbench will be compared with the ones obtained by CFD simulations with moving wall (figure 1).

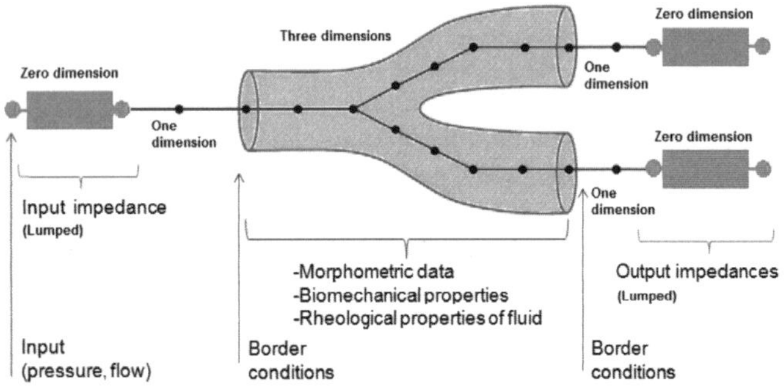

Figure 1. In-silico characterization by FSI+CFD. Coupling of 3D, 1D, and 0D models to perform a fluid-structure interaction analysis of the flow and wall motion in an arterial bifurcation.

The main purpose of the present paper is to describe the development of a new multi-parameter measurement system for an in-vitro characterization of the biomechanics and hemodynamics in arterial bifurcations. This set-up represents an improvement relative to previous set-ups developed by the group FCIEN-FMED and is presently under development. It poses several interesting problems related with signal acquisition using a variety of sensors. Some representative experimental results are shown and briefly discussed.

2. Hemodynamic work bench simulator (HWBS)

The HBWS closed circuit is composed by a pump, the tubing between the pump exit and the entrance to the arterial bifurcation, the arterial bifurcation being studied, tubing between each branch of the bifurcation exit and the entrance to a fluid reservoir, adjustable constrictions on each branch, a fluid reservoir with mean pressure adjustment, and a returning tubing from the reservoir to the pump (figure 2).

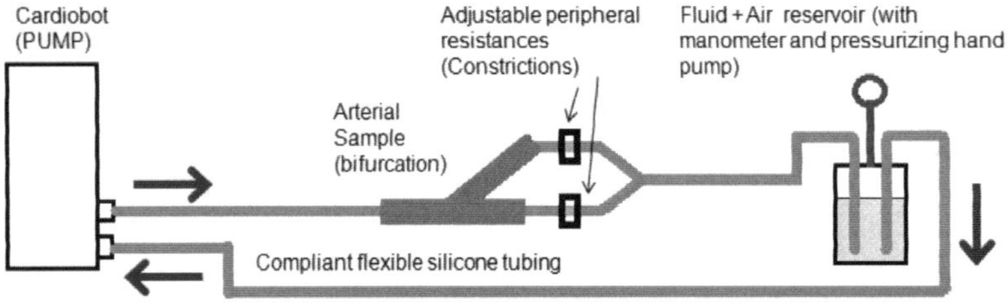

Figure 2. HWBS fluid circuit schematic diagram. Compliant silicone tubing is used for the circuit.

An artificial heart (Cardiobot) is used as the pulsatile pump, with frequency and ejection volume adjusted at demand. Cardiobot is an electronically controlled programmable pump, specially developed at LAU-FCIEN-UdelaR [6] in order to generate different flow and pressure patterns in the HWBS. Cardiobot generates a trigger output at the beginning of each systole. This signal is used as a time reference and for triggering.

The output load impedance distal to each branch of the arterial bifurcation can be adjusted by means of variable restrictions (variable clamp). For each branch, there is an independent restriction located far away from the corresponding branch output. Each restriction introduces localized mechanical impedance mismatch that produces reflected waves. The combination of upstream and downstream pressure pulses produce global pressure patterns in the arterial sample. Adjusting each restriction (amongst others bench parameters), near-physiological pressure pulse and flow patterns can be obtained in the bifurcation.

In each experimental configuration, the artificial heart is operated at constant frequency. The measurements are done after a steady regime of pulsatile flow is attained in the circuit. The already established dynamic regime and the trigger signal allow relocation and exchange of sensors during measurements (for example, A-Scan and Pulsed Doppler measurements can be done in the same region alternating probe placement). The samples are mounted on an arterial bifurcation fixing system (couplers and adjustable arms) and submerged in a physiological solution pool. This pool is rounded by an external water pool where temperature control can be done (figure 3).

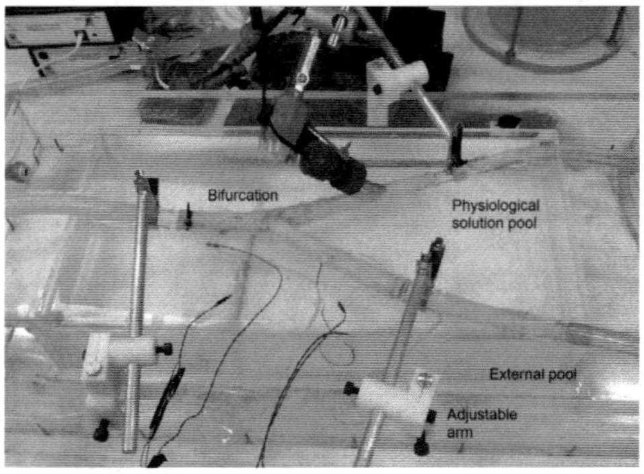

Figure 3. Photograph of arterial bifurcation in the physiological solution pool. Figure shows 3 adjustable arms, pressure and diameter sensors and pulsed Doppler probe.

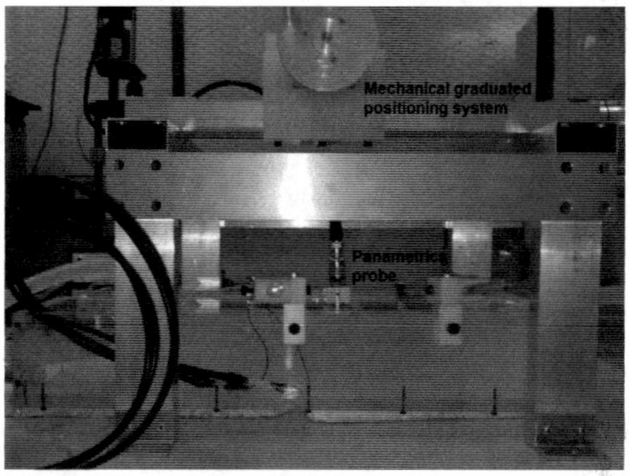

Figure 4. Photograph shows mechanical graduated positioning system (in x, y, z axes) and Panametrics V313 probe during morphometric measurements.

The circuit can be filled with physiological solution or blood treated with EDTA sodic (rheological properties previously obtained with Brookfield, LVDT –II+ plate-cone viscometer). Temperature control can be done in the reservoir and in the external pool, with a set-up point of 37°C (measured using one thermistor in each of the above mentioned recipients). The reservoir is a glass flask containing blood or physiological solution and air. It's mean pressure can be adjusted at constant values by means of a hand pump and a differential pressure sensor. The HWBS can be operated with arterial samples, phantoms or other man-made tubes or bifurcations under static and dynamic conditions.

Other experimental setups are designed and operated exclusively to characterize mechanical properties of the arterial or venous walls, but can't be used to study wave propagation and fluid-structure interactions [7]. Bertram et al. [8] developed an experimental setup for arterial or tube segments that allow pulse propagation measurements under near physiological pressure pulse and flow patterns. Their setup employs rigid wall tubing. Two closed fluid compliance reservoirs (with a pressurized air chamber inside) are located before and after the sample. The first one attenuates pressure transients originated in the closure of the valves of the piston pump. The second one is intended to allow the modification of mechanical impedance. A Darcy type hydraulic restriction located after the second chamber introduces a linear loss of charge. However it can't be adjusted during operation. Our setup allows the adjustment of the two restrictions independently and during operation.

3. Description an discussion of system components and their interactions

The sample is positioned in the bifurcation fixing system trying to maintain the original geometry of the sample (lengths and angles measured during sample extraction surgery). Then, we fill the circuit with the fluid and purge the bubbles. With no flow and no transmural pressure in the circuit, measurements are done over the bifurcations in A-Scan mode in order to obtain lengths, internal and external diameters and wall thickness (morphometric data useful to construct a simplified finite-element grid for a CFD model). A 15MHz Panametrics V313 probe fixed on a mechanical graduated (1,5mm resolution) positioning system is connected to a Lecoeur A-Scan module controlled by a PC running correlation algorithm developed to obtain the measurements (figure 4).

After that, several hemodynamic conditions are simulated and measured (figure 5), adjusting the flow pattern of the Cardiobot, reservoir pressure and the variable restrictions (each one far enough from the output of each bifurcation branch, in order to obtain near-physiological flow and pressure patterns in the sample). The non-linear characteristics of this kind of restrictions are similar to those imposed by cardiac valve stenoses [9]. External diameter measurements are taken at suitable points in each segment of the arterial bifurcation using a gold-standard System 6 Mainframe equipped with sonomicrometer ultrasonic modules [9] and ultra resolution techniques [4]. Sonomicrometer ultrasonic crystals (5 MHz, 2 and 3mm diameter) were handcrafted and fixed to Dacron patches. These crystals are sutured to the outer face of the arterial wall (adventitia) in opposite positions relative to a diameter. Pressure waveforms are obtained using Königsberg sensors P2,5-S with high frequency response (1kHz) located near diameter measurement points. Sensors are sutured sensing surface points to the lumen in order to obtain the hydrostatic component of the pressure. Velocity profiles at different locations are obtained using a multigate Doppler system (DOP, Signal Processing SA). Micro-spheres powder (scatterers) are used when the HWBS circuit is filled with a physiological solution or water-glycerine mixture fluid. Cross sectional flow is measured using suitable diameter perivascular transit time flowmeters (Triton flowmeters).

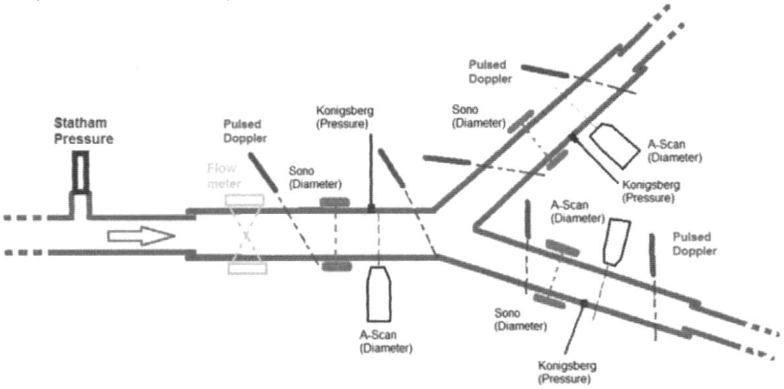

Figure 5. Schematic diagram of sensor placement.

A Statham pressure sensor (located in the tubing near the common segment) is used for monitoring and initial adjustments procedures. In order to give a time reference for the measurements of the different sensors a periodic rectangular pulse signal is generated by the Cardiobot related with pulsatile pump function. Signals are acquired at 3kHz using 12-bit data acquisition modules (NI USB-6009 and LabJack U3-HV) operated from a specially developed Matlab application. Two Tektronix TDS2024B,

100MHz-4 channel digital oscilloscopes are used for initial adjustments and for monitoring purposes. After all the sensors are placed (and after temperature stabilization), gain and offset adjustment of modules is done. Voltage vs. diameter and voltage vs. pressure curves are obtained (pressure sensors are calibrated against a mercury manometer with static measurements at discrete pressure values). Estimation of propagation velocity of sound in the sample wall and wall thickness can be done using an auxiliary acoustic mirror located at a known distance from the Panametrics probe and with the sample between them. These parameters are loaded in the A-Scan developed processing software. Pressure and diameter samples are processed on-line in a computer running a specific developed algorithm. Biomechanical properties of the wall are obtained by means of model parameter adjustment [10].

The generalized static and dynamic characteristics of the sensors [11] used in the new system were checked. To validate the experimental results obtained using the new multi-parameter measurement system, we studied the performance and interoperability of each sensor in the framework of the whole measurement system, taking into account possible external influences. A survey of possible external influences was done. The influence of noises of the same physical nature as the input signals to the sensors and the EMC aspects were studied. For example, sonomicrometer measurements are sensitive to external radiated RF perturbations. These perturbations generated spurious triggering at the RF processing stage, altering the stability of the measured distances. This was corrected by the use of special shielded cable arrangements. Other radiated and conducted electromagnetic interference was reduced by the implementation of a shield to isolate the workbench from its external environment. Table 1 shows the main interfering and modifying inputs and their possible compensation, both between environment and HWBS system and between HWBS subsystems. As an example of interfering input, let us consider transit time flow measurement. Also, let us consider that transit time flowmeter internal diameter is slightly bigger that external arterial diameter. Pulsatile flow causes vessel diameter variation at the cross sectional flow measurement points. This diameter variation is a interfering input for the flow meter. Transit time flow meter calculates cross sectional flow Q(t) by measuring upstream T_{upstr} and downstream $T_{downstr}$ propagation times [11,12]. The voltage output V_{out} depends on the time difference, system gain χ and an error δT:

$$V_{out}=\chi\cdot(T_{upstr}-T_{downstr})+\chi\cdot\delta T \tag{1}$$

δT is mainly of electronic origin and $\chi\cdot\delta T$ can be at least partially compensated. The time difference verifies:

$$(T_{upstr}-T_{downstr})=K\bullet\left(\frac{Q(t)}{R_i(t)}\right) \tag{2}$$

K depends of internal parameters of the module, speed of sound and nature of flow (laminar or turbulent). So, neglecting δT, V_{out} is proportional to $\frac{1}{R_i(t)}$ and Q(t). The internal radius of the vessel $R_i(t)$is an interfering input that modulates the flow signal. In general, $\left|\frac{\delta R_i(t)}{\overline{R_i}}\right|\leq 0{,}1$ so an error of up to ±10% can be expected in flow measurement. This error can be compensated in post processing if internal diameter is measured at the same region and same time (assuming a circular cross section vessel). As an example of modifying inputs, let us consider A-Scan measurement. Environmental and laboratory EMC interferences are modifying inputs that can be compensated by the nature of the processing algorithms (averaging and cross correlation of echo signals). Other sources of interfering and modifying inputs and their possible compensation techniques can be found in table 1.

Before operating the system with arterial samples, suitable phantom models where used for sensor calibration and error estimation. The biomechanical properties of the arterial segment obtained by parameter estimation will be employed in the model for the numerical simulation of the interaction between structure and fluid. Velocity profiles obtained in the workbench will be compared with the ones obtained by CFD simulations with moving wall.

Table 1. Main interfering and modifying inputs and their possible compensation.

Subsystem	Main Interfering input	Possible Compensation techniques	Main Modifying input	Possible Compensation techniques
Sonomicrometer	US waves of other subsystems	Spatial disposition of subsystem sensors	Cardiobot power stage interferences	Special coaxial cable configuration for receiver. Sonomicrometer trigger level adjustment
Sonomicrometer	US waves of other subsystems	Non-simultaneous operation (only for nontransient operation)	Other laboratory and environmental electrical interferences	Metal wire mesh shielding, dedicated online double conversion UPS with isolation transformer. Sonomicrometer trigger level adjustment
A-Scan	US waves of other subsystems	Spatial disposition of subsystem sensors	Cardiobot power stage interferences	Echo Signal averaging. Cross correlation processing
A-Scan	US waves of other subsystems	Non-simultaneous operation (only for nontransient operation)	Other laboratory and environmental electrical interferences	Echo Signal averaging. Cross correlation processing
A-Scan	Possible interfering spurious vibrations	Echo Signal averaging. Cross correlation processing	–	–
Flow meter	Variable vessel diameter in time	Opposing input (generated from simultaneous measurement of vessel diameter)	Other laboratory and environmental electrical interferences	Metal wire mesh shielding, dedicated online double conversion UPS with isolation transformer.
Flow meter	Difference between sensor and vessel cross sectional area	Choose suitable sensor diam. or correction factor (if above opp input not applied)	Internal and external noise	Output filter (100, 30, 10Hz)
Pressure (Sthatham)	Trapped air bubbles	Purging the catheter	–	–
Pressure (Sthatham)	Mechanical vibrations/movement catheter	Short catheter. Avoid movements during measurements	–	–
Pressure (Konigsberg)	Mechanical vibrations/movement catheter	Avoid tip shocks during measurements	–	–
Pressure (Konigsberg)	Temperature variations	Correction of measurements (known offset in mmHg/^{0}C)	–	–
Pulsed Doppler	Vessel movements (lateral)	Minimizing vesel movements. Use transducer array (Open System Lecoeur)	–	–

4. Examples of data obtained and post-processing

Static and dynamic measurements of diameters (external and internal), pressures, flows and velocity profiles were done in several locations in the common segment and branches of the bifurcations. Figure 6 shows the simultaneous acquisition of diameter, pressure and cross sectional flow (without filtering) for a location in the common segment as well as the trigger signal generated by Cardiobot pump (rising edge corresponds to systole beginning). Figure 7 shows velocity profiles (without filtering) obtained by pulsed Doppler, in the same place. Vessel pulsation can be seen through the variation of the profiles width over the x-axis (distance in mm).

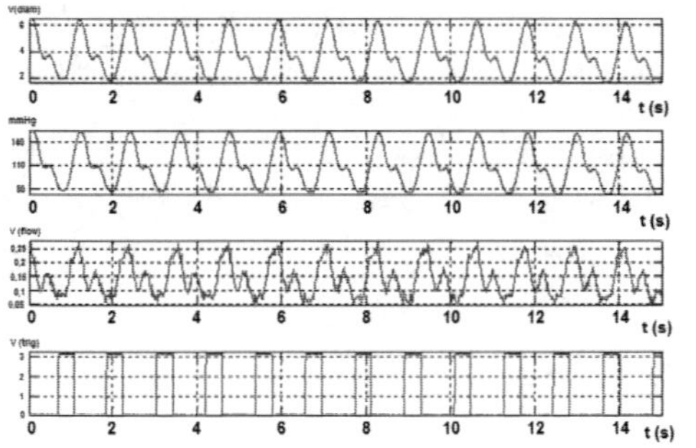

Figure 6. Diameter, pressure and flow in the same location for a main branch. Trigger signal can be seen at the bottom.

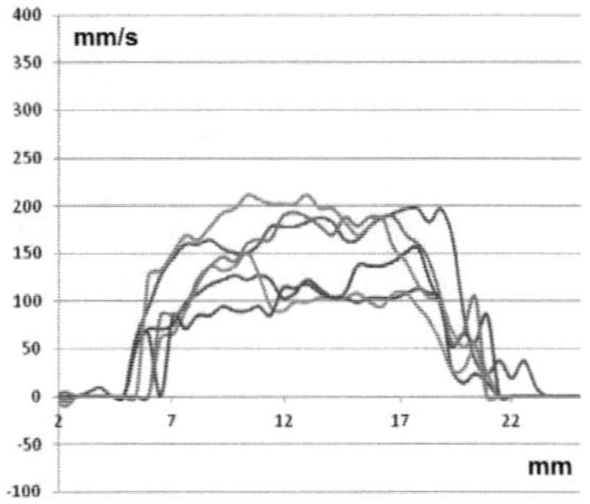

Figure 7. Pulsed Doppler velocity profiles (mm/s) obtained for a main branch. Vessel pulsation can be seen through the variation of the profiles width over the x-axis (distance in mm).

Figure 8 shows diameter and wall thickness variation during a slow depressurization experiment suitable to estimate relaxed (quasi-static) elasticity modulus (figure 9). Measurements are obtained from A-Scan signals processed with a specially developed acquisition and correlation algorithm [5].

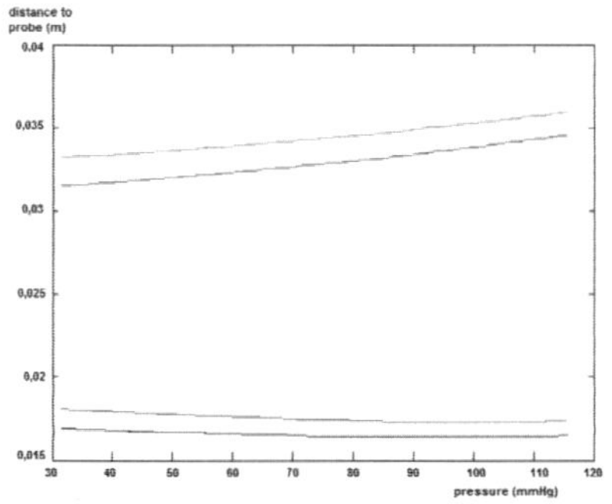

Figure 8. Internal and external diameters (measured with cross correlation techniques) vs. transmural pressure for a given cross section of the sample, during a slow depressurization experiment (from right to left).

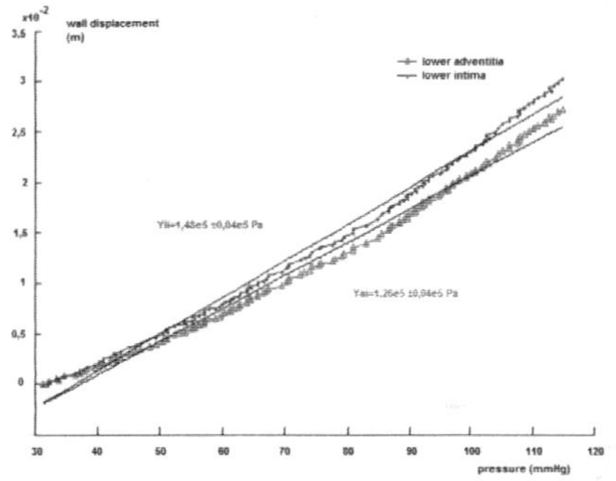

Figure 9. Young modulus estimation for a given cross section of the sample.

5. Conclusions

Near-physiological flow and pressure patterns can be obtained for different kind of arterial bifurcations (specially for carotid bifurcations) by means of adjusting compliant tube lengths and hydraulic resistances as well as the Cardiobot pulsation patterns. This kind of flow and pressure patterns are not strictly necessary if the intention is a biomechanical characterization of the arterial wall. However, realistic flow and pressure patterns are necessary to study fluid-structure interactions in scenarios as near as possible to in-vivo ones.

Working in a steady pulsation regime with a time reference signal allowed us to take different physical measurements in the same point of the arterial sample locating there, at different time instants, different sensors and relating the signals recorded to the same phase of the periodic regime. This allowed us to overcome the difficulties due to the size of and interference between the different sensors. Of course, the duration of the whole experiment must be short enough so that the properties of the arterial bifurcation don't change significantly.

Although the static and dynamic characteristics of the sensors used were suitable for the attaining of the posed objectives, when the whole system was assembled, several modifying and interfering inputs appeared due to intra-system interference and system-environment interactions. Compensation techniques applied worked successfully.

The obtained measurements can be used for: 1-Mesh construction for CFD modeling using morphometric arterial sample data obtained by caliper measurements and ultrasonic measurements techniques (US probe fixed to the mechanical graduated positioning system shown in Fig.4). 2-Parameter estimation of the biomechanical properties of the arterial wall, suitable to be employed in a model for the numerical simulation of the interaction between structure and fluid. 3-Assessment of velocity profiles in order to make a comparison with the ones obtained by CFD simulations of the arterial carotid bifurcation sample, taking into account the movement of the arterial wall. Estimation of the endothelial wall shear stress for a carotid arterial bifurcation at different points, under a periodic pulsated flow can be done by CFD simulation. 4-Establish boundary conditions (pressure and cross sectional flow) at the entrance and at the exits of the arterial sample, for CFD-FSI numerical calculations.

6. References

[1] Brum J, Bia D, Benech N, Balay G, Armentano RL and Negreira C 2010 Setup of a cardiovascular simulator: application to the evaluation of the dynamical behavior of atheroma plaques in human arteries *Physics Procedia.* **3** 1095-1101

[2] Balay G, Brum J, Bia D, Armentano RL and Negreira CA 2010 Improvement of artery radii determination with single ultra sound channel hardware & in vitro artificial heart system *Conf. Proc. IEEE Eng. Med. Biol. Soc.* 2521-24

[3] Bia D, Armentano RL, Zocalo Y, Barmak W, Migliaro E and Cabrera Fischer EI 2005 In vitro model to study arterial wall dymamics through pressure-diameter relationship analysis *Latin American Applied Research* **35** 217-225

[4] Nichols W, O'Rourke M and Vlachopoulos Ch 2011 McDonald's blood flow in arteries: theoretical, experimental and clinical principles *Hodder-Arnold*

[5] Armentano RL, Bia D and Cabrera Fischer E 2006 An in-vitro study of cryopreserved and fresh human arteries: a comparison with ePTFE prostheses and human arteries studied non-invasively in vivo *Cryobiology* **52(1)** 17-26

[6] Balay G 2012 Elasticity in arterial tissues, design of an in-vitro artificial heart and a new ultrasonic method of arterial elesticity assessment *M.Sc. thesis (Physics), LAU-FCIEN-Ude la R*

[7] Hayashi K, Mori K and Miyazaki H 2003 Biomechanical response of femoral vein to chronic elevation of blood pressure in rabbits *Am J Physiol* **284** H511–18.

[8] Bertram CD, Pythoud F, Stergiopulos N and Meister JJ 1999 Pulse wave attenuation measurement by linear and nonlinear methods in nonlinearly elastic tubes *Medical Engineering and Physics* **21** 155–166.

[9] Chandran K, Rittgers S and Yoganathan A 2012 *Biofluid mechanics: the human circulation* **2nd ed** CRC Press

[10] Gamero LG, Armentano RL, Barra JG, Simon A and Levenson J 2001 Identification of arterial wall dynamics in conscious dogs *Exp Physiol* **86** 519-528

[11] Webster J 2010 *Medical instrumentation: application and design* J. Wiley & Sons Inc **4th ed** NJ chapter 1, 19-25

[12] *System 6 Manual* 1997 Triton Technology Inc

Acknowledgements

The first author acknowledges ANII for grant BE_POS_2010_1_2588 and PEDECIBA for support for research.

Electrodes for bio-application: recording and stimulation

M B A Fontes[1]
Electronics Systems Department of Sao Paulo College of Technology DSE/FATEC-SP/CEETEPS, 01124-060, Sao Paulo, SP, Brazil

E-mail: bariatto@fatecsp.br

Abstract. Recording and stimulation electrodes applied on excitable tissue are the basis of electrophysiological research, such as brain, muscles, peripheral nerves or sensory systems. Electrode-electrolyte impedance is one of the important characteristics due to its influence on the signal/noise ratio, signal distortion and built-up voltage. Strategies to lowering and tuning the impedance are achieved by biasing iridium oxide modified platinum microelectrodes. Surface and impedance analysis after pulse stimulation are also addressed.

1. Introduction

The bioelectric potentials carried in electrolytic media in form of ionic currents are transduced by the electrodes into electronic signals. This takes place by means of capacitive coupling (without net charge transfer), as it happens during recording, and by charges transfer reactions in which the electrodes and ions in the physiologic environment exchange electrons through redox reactions. This process occurs mainly during stimulation. Those reactions can be reversible or irreversible, usually generating gases at the electrode site (H_2, O_2, C_{l2}) and leading to damaging/poisoning of the tissue and electrode corrosion due to its oxidation [1-2].

Strategies to lower the impedance are of key importance in applications such as neural signals recordings. Iridium Oxide (IrO_2), a biocompatible metallic oxide, is an excellent candidate to accomplish this impedance reduction given its texture. Iridium Oxide is of particular importance in cases where electrical stimulation is required [3,4] because it is an electroactive material, i.e., it forms a fast-kinetics redox system on the electrode surface, therefore allowing Faradaic charge transfer. It is also known to exhibit electrochromic behavior, i.e. a reversible change of its color in response to the application of an electrical charge. Electrochromism is also followed by a change in impedance, whose value is decreased for anodic potentials [5]. Once the impedance is set, it will stay at that value until a new potential is applied. The process is fully reversible within a time scale of milliseconds.

2. Electrode Impedance

In the recording/stimulation of bioelectric events, the electrode performs the function of transduction from the ionic currents in the electrolyte to an electric current in the measurement system. The electric contact between an electrode and living tissue has an electric impedance whose high values reduce signal to noise ratio and increase signal distortion. This is particularly relevant for microelectrodes due to their reduced dimensions and also because their impedance has the tendency to increase even more due to adsorption of biological material [6,7].

Simulations based on simple Randles model show, Fig. 1, that for microscale electrodes, the

[1] To whom any correspondence should be addressed.
Published under licence by IOP Publishing Ltd

capacitive coupling is the major component for most physiological signals (mHz-kHz) and the electrode impedance and thermal noise (Johnson-Nyquist) increase as the electrode area decreases [8-10].

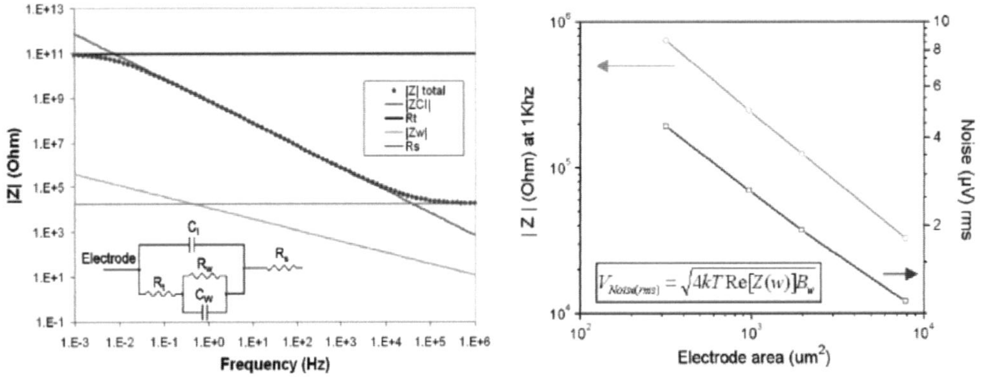

Figure 1. Platinum electrode (Φ=20μm) impedance simulation (left) and thermal noise at 1 KHz (right), based on Randles model.

3. Electrode Modification

IrO_2 coatings have been obtained by directly electroplating over platinum electrodes. The electroplating solution is water based containing iridium chloride ($IrCl_4.H_2O$), hydrogen peroxide (H_2O_2), oxalic acid $(COOH)_2.2H_2O$ and potassium carbonate (K_2CO_3) to adjust the pH to 10.5 [11]. Sample pre-treatment is performed through surface degreasing with acetone followed by isopropanol and rinsing in deionized water. A preliminary treatment to improve adhesion can be done by slightly etching the platinum surface in *aqua regia* solution (3 HCl : 1 HNO_3 : 2 H_2O) at 50°C, or by using the standard RCA organic silicon cleaning (1 NH_4OH: 1 H_2O_2 : 5 H_2O) at 70°C.

The use of electroplated IrO_2 has the advantage of attaining deposition on various electrode geometries and morphologies (cavities), easy control of process (hence material properties) and intrinsically low cost, especially if compared with sputtering. Furthermore, preliminary experiments show enhanced impedance tuneability in comparison with sputtered iridium oxide. On the other hand electroplating needs a base conductive layer, usually platinum (Pt), although this can be overcome by using electroless or autocatalytic deposition processes.

Several electroplating protocols applying potential control (potentiostatic) or current control (galvanostatic) were used [12]. We have utilized a CompactStat Electrochemical Analyzer (Ivium Technologies) to realize such depositions and to perform Electrochemical Impedance Spectroscopy (EIS) characterization of the obtained electrodes. We have used the standard three-electrode method in both procedures: Pt/IrO_2 (i.e. the electrode to be modified) as the working electrode, an Ag-AgCl reference electrode and a stainless steel cup as the auxiliary electrode. All EIS characterization was performed in physiological solution (0.9% NaCl, pH=7.4) except for the in situ monitoring where the deposition solution was used instead.

We monitored the impedance evolution during the deposition, scanning at a very slow rate the voltage (1mV/s for potentiostatic deposition) or the current (10μA/s in galvanostatic deposition), Fig. 2. We observed a drop in the impedance at 10 Hz when the voltage reaches 0.6V. The same behavior can be obtained at a current density of 2.5 A/m^2 The impedance does not change within the layer thickness. This is because the surface roughness stays approximately the same for as long as the deposition conditions are not changed.

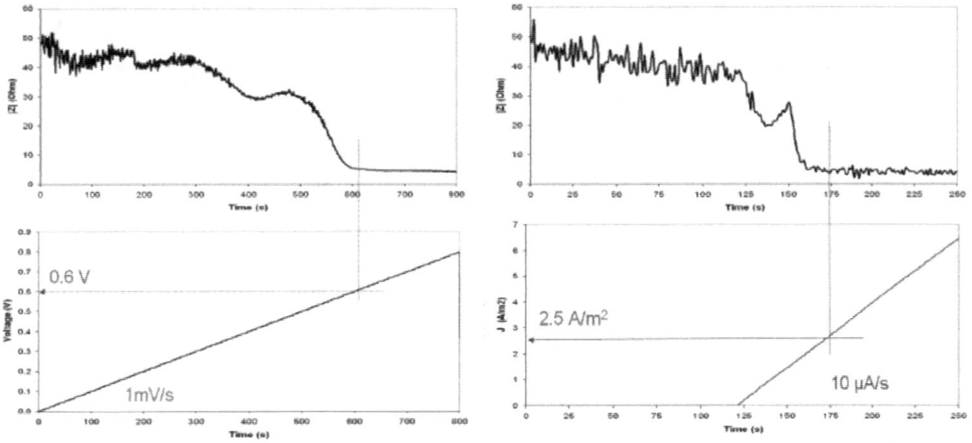

Figure 2. In situ potentiostatic (left) and galvanostatic (right) deposition with impedance measurements over platinum sample A=2cm^2

A series of galvanostatic depositions at a constant current (1 A/cm^2) and at different times lead to the conclusion that indeed the impedance drops right away when the first layer covers the platinum underneath, Fig. 3.

SEM analysis showed that depositions beyond a thickness of 200 nm leads to film cracks and further impedance decreasing due, probably, to exposition of the underneath platinum layer. Below that thickness, very good uniformity was achieved. AFM analysis indicated a surface roughness around 8 times greater than that for the base platinum electrode.

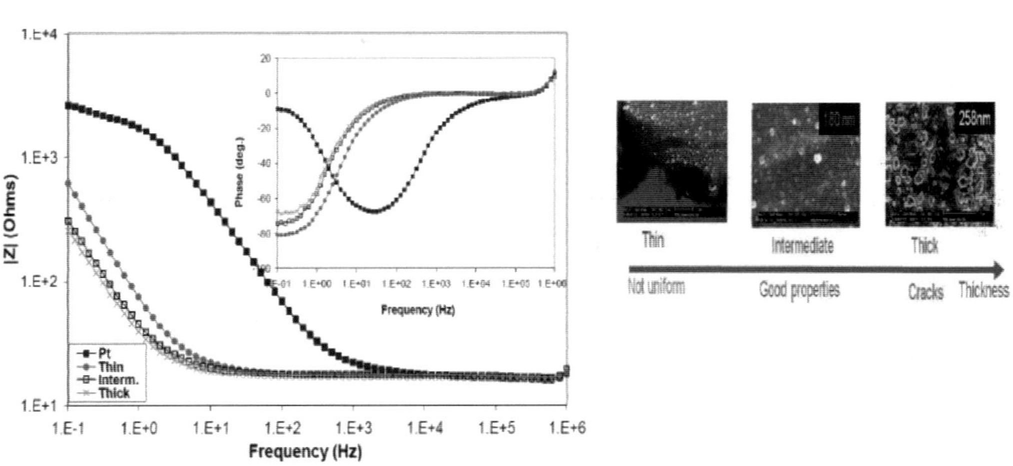

Figure 3. Galvanostatic deposition at constant current showing changes in the impedance spectrum (left) and the time evolution (right) of iridium oxide plating on platinum samples with 2 cm^2 area.

Using the known voltage and current range, several depositions schemes were performed, including constant current and constant voltage, over platinum microelectrodes (Φ=100μm). Constant current deposition leads to a similar decreases on the impedance, evaluated at 1 KHz – about 14 times -but has better performance over a large frequency range, Fig. 4.

Simulation of the modified electrode was achieved by modifying the previously reported Randles model for a platinum electrode, with a large capacitor in parallel, resembling the additional active area (200x) provided from the IrO_2 film, also in Fig. 4.

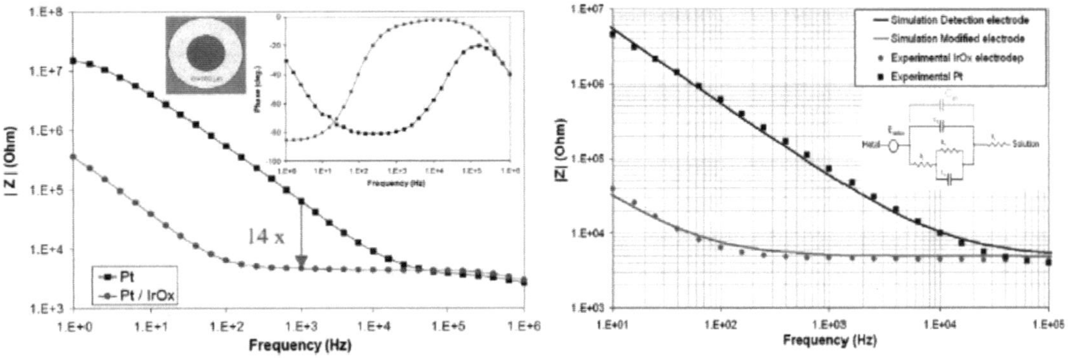

Figure 4. Impedance change after galvanostatic deposition of IrO_2 (left) and simulations and experimental data with and without IrO_x covering a platinum microelectrode (right) with $\Phi=100\mu m$ and $50\mu m$, respectively.

4. Impedance Tuning

Electrochemical Impedance Spectroscopy (EIS) measurements are usually performed at equilibrium potential (E0) or at zero volts, but this value can be adjusted as a biasing to the superimposed AC signal [8].

In order to analyze how the electrode area influences the impedance tuning effect, an array of platinum microelectrodes with diameters ranging from 10 µm to 100 µm were electroplated with IrO_2. No effect is observed for platinum electrodes when EIS is performed with a bias voltage ranging from 0 to 1 V in 100 mV steps, Fig. 5. On the other hand, platinum electrodes plated with IrO_2 show bias dependent impedances for voltages above 500 mV, where impedance reduction up to two orders of magnitude could be achieved for 20 μm diameter electrode -@1 kHz , 1V -Fig. 6. at 1KHz, 1V-Fig. 6.

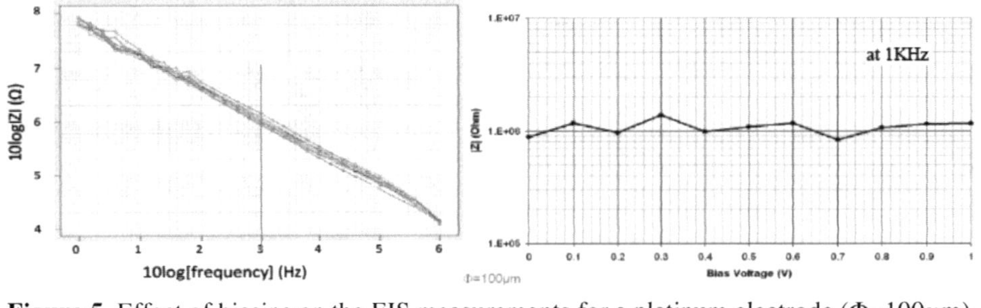

Figure 5. Effect of biasing on the EIS measurements for a platinum electrode ($\Phi=100\mu m$).

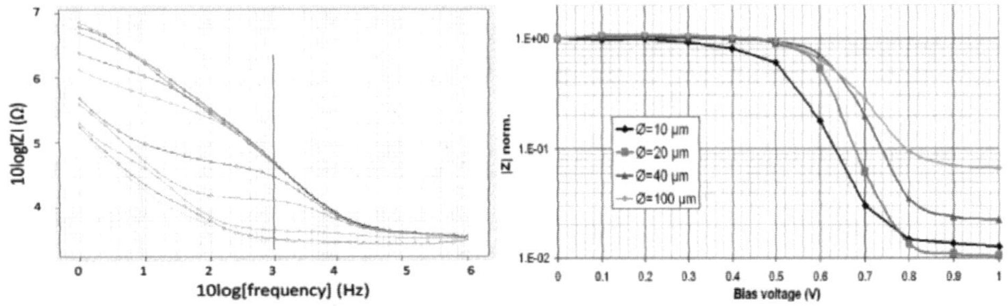

Figure 6. Effect of biasing on the EIS measurements of IrO_2 coated platinum electrodes: EIS scans at various bias voltages on a 100 μm diameter (left) and impedance versus bias voltage plots for various sample diameters (right).

Analyzing the biasing dynamic behaviour, we applied a 1V pulse train (width=5 s) to the platinum electrodes where no changes in impedance were observed, as expected, Fig. 7 (left). On the other hand, applying the same pulse train to IrO_2 coated electrodes, a reproducible and reversible impedance decrease was observed within a time response of less than 200 ms, Fig. 7 (right).

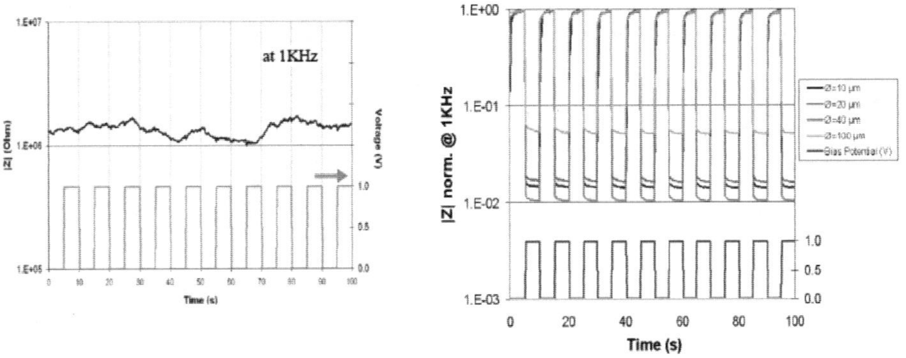

Figure 7. Effect of applying a pulsed bias on impedance measurements for platinum, Φ=100μm, (left) and platinum modified IrO_2 (right) microelectrodes.

5. Electrode Stimulation

The aim of the stimulation is to provide enough charge to trigger an action potential at a given spot. For example, at the muscle, periferical nerve and cortical surface level requires charges up to 0.2 to 5 mC per stimulation pulse and intracortical region needs only 8 to 64 nC per stimulation pulse [3,4].

As the electrode area goes down, the amount of charge density increases leading to a failure of the electrode. Optimizing a parameter called Charge Store Capacity (CSC) provides more charge per electrode area, therefore prevent the electrode failure extending its usable life. This parameter can be extracted through the Cyclic Voltammetry (CV) wave, where the cathodic region (negative current) is related to the amount of electrons that the electrode can supply at a given range of voltage [1], Fig. 8 (left).

Stimulation protocols are usually carried out in form of pulses of voltage or current, where the last one is more common. Several combinations can be used but charge balanced biphasic protocols are usually standard due to the fact that the net charge is zero [2], Fig. 8 (right).

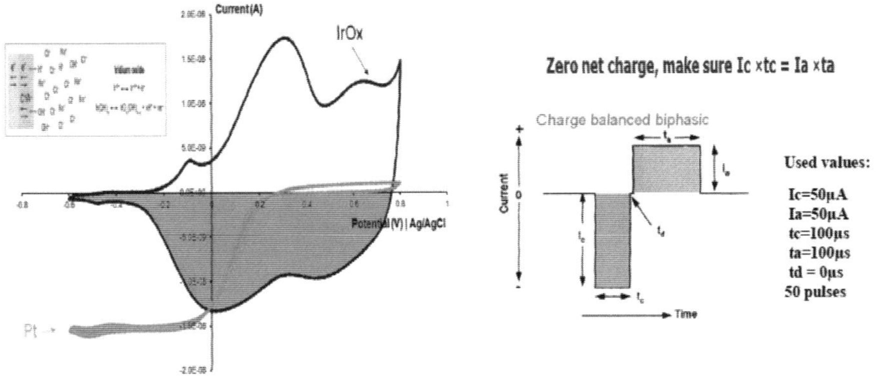

Figure 8. Cyclic voltammetry indicating the Charge Store Capacity (CSC) -filled area -for IrO_2 and Pt electrodes (Φ=40μm) in saline solution (left) and balanced biphasic stimulation pulse protocol parameters (right).

Before and after stimulation, cyclic voltametry and impedance of the electrodes were performed. Results showed an improvement for platinum electrodes, most probably due to the electrode cleaning. Iridum oxide coverage presents a decreasing of their CSC with decreasing impedance. Changes in the electrode surface, Fig. 9, were observed optically.

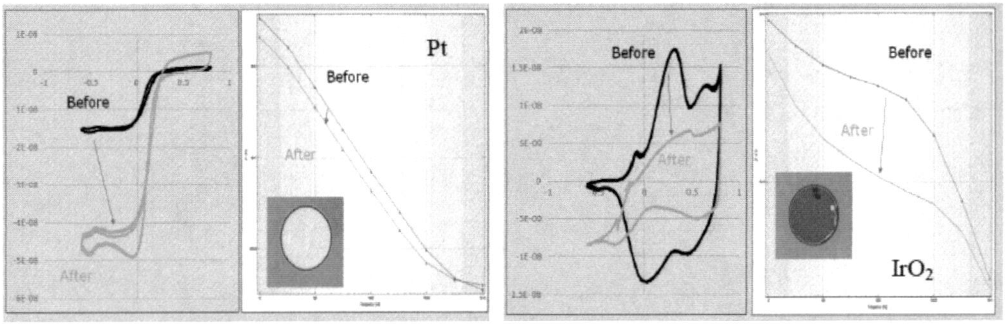

Figure 9. Cyclic voltametry and impedance measurements, before and after electrodes stimulation (Φ=40μm), for platinum (left) and iridum oxide (right).

One of the possible causes of the electrode surface modification or even its failure, during stimulation, is associated to the irreversible reactions described before, which happens outside of the water hydrolysis window (-1V to +1V against Ag/AgCl reference electrode) that generates oxygen and hydrogen, respectively. This building up voltage is minimized by low impedance values of the electrodes. All stimulated electrodes showed, at some extend, voltages on that region, indicating possible long-term degradation. This is particularly valid for the platinum electrode, where most of the time stays at the unsafe zone, compared with iridium oxide, Fig. 10.

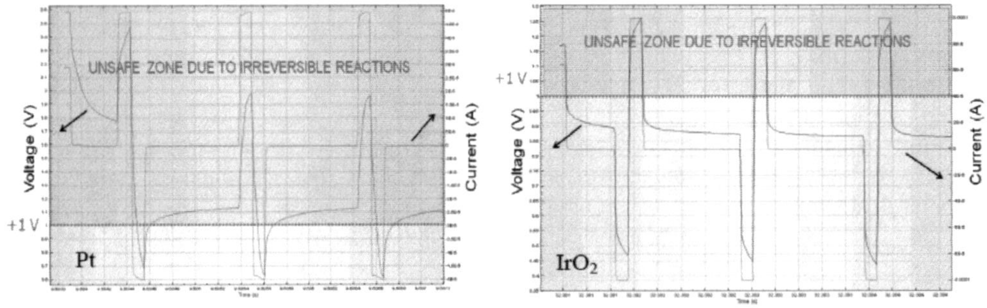

Figure 10. Applied current and recorded voltage during the stimulation for all electrode materials (Φ=40μm), for platinum (left) and iridium oxide (right).

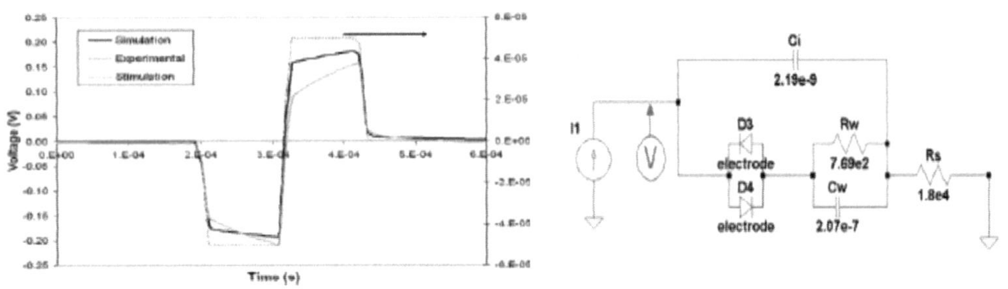

Figure 11. Large signal model of the stimulation for a platinum electrode (Φ=40μm).

Preliminary simulation for the pulses could be performed using the same Randles small signal model used for impedance, changing the transference resistance to a diode model that resembles better the large current due to the faradic reactions, Fig. 11.

6. Conclusion

Modification of platinum microelectrodes using a biocompatible metallic oxide film -Iridium Oxide - successfully decreased their impedance by a factor of 14 at 1KHz. *In situ* Electrochemical Impedance Spectroscopy allowed determine the deposition window either for potentiostatic or galvanostatic processes. SEM images showed film cracks over 200nm thickness due to mechanical stress. A modified electrode-electrolyte model (Randles circuit), suggested an active area increasing over 200 times.

Impedance tuning of iridium oxide electrodes has been demonstrated. By applying a DC voltage between 0 and 1 V it is possible to vary electrode impedance by as much as two orders of magnitude, therefore enabling the electrode impedance to be electrically programmed. The process is fully reversible with a time response in the millisecond range. Iridium Oxide electrodes combines a low impedance and high charge storage capacity, therefore it is a superior candidate for recording and stimulation.

Acknowledgments

The author wish to acknowledge the Interuniversity Microelectronics Center (IMEC) and Division of Neurophysiology at Catholic University of Leuven both in Belgium and also the Department of Microsystems Engineering (IMTEK), University of Freiburg, Germany.

The work described in this paper was performed as part of the Information Society Technologies (IST) Integrated Project NeuroProbes of the 6th Framework Program (FP6) funded in part by the European Commission (Project number IST-027017).

Finally the author also acknowledges the CNPq for the grant (201720/2007-8).

7. References

[1] Frölich R, Rzany A, Riedmüller J, Bolz A and Schaldach M 1996 Electroactive coating of stimulating electrodes *J. Mat. Sci. Materials in Medicine* **7** 393-397

[2] Weiland J D and Anderson D J 2000 Chronic neural stimulation with thin-film, iridium oxide electrodes *IEEE Trans. Biomed. Eng.* 47(7) 911-918

[3] Meyer R D, Cogan S F, Nguyen T H and David Rauh R 2001 Electrodeposited Iridium Oxide for Neural Stimulation and Recording Electrodes *IEEE Trans. Neural Systems Rehab. Eng.* 9(1) 2-11

[4] Weiland J D and Anderson D J 2000 Chronic neural stimulation with thin film, iridium oxide stimulating electrodes *IEEE Trans. Biomed. Eng.* **47** 911–918

[5] Pauporte R and Durand R 2000 Impedance spectroscopy study of electrochromism in sputtered iridium oxide films *J. Appl. Electrochem.* **30** 35-41

[6] Kovacs G T A 1994 *Microelectrode models for neural interfaces," in Enabling Technologies for Cultured Neural Networks* (D.A. Stenger and T.M. McKenna, Eds.), pp. 121-165, Academic Press, New York

[7] Franks W, Schenker I, Schmutz P and A. Hierlemann A 2005 Impedance characterization and modeling of electrodes for biomedical applications *IEEE Trans. Biomed. Engin.* **52(7)** 1295-1302

[8] Macdonald J R (editor) 1987 *Impedance Spectroscopy: Emphasizing Solid Materials and Systems* (John Wiley & Sons)

[9] David A. Borkholder 1998 Cell Based Biosensors Using Microelectrodes, PhD. Dissertation, Stanford University

[10] Lempka S F et al 2006 Optimization of Microelectrode Design for Cortical Recording Based on Thermal Noise Considerations EMBS Annual International Conference (New York City, USA, Aug 30-Sept 3, 2006)

[11] Yamanaka K 1989 Anodically electrodeposited iridium oxide films (AEIROFs) from alkaline solutions for electrochromic display devices *Jpn. J. Appl. Phys.* **28** 632-637

[12] Fontes M B A 2010 In Situ Impedance Monitoring of Iridium Oxide Electrodeposition for Neural Recording *Proc. 7th IberoAmerican Sensors Conference* (Nov 9-11, Lisbon, Portugal)

Electrochemical magneto-immunosensing of *Salmonella* based on nano and micro-sized magnetic particles

D Brandão[1], S Liébana[1], S Campoy[2], P Cortés[2], S Alegret[1], M I Pividori[1,3]
[1]Grup de Sensors i Biosensors, Unitat de Química Analítica, Universitat Autònoma de Barcelona, 08193, Bellaterra, Barcelona.
[2]Grup de Microbiologia Molecular, Departament de Genètica i de Microbiologia, Universitat Autònoma de Barcelona, Bellaterra, Spain

E-mail: isabel.pividori@uab.cat

Abstract. A very simple and rapid method for the detection of *S. enterica* is reported. In this approach, the bacteria were captured and preconcentrated with magnetic particles through an immunological reaction. A second polyclonal antibody labeled with peroxidase was used for the electrochemical immunosensing based on a magneto-electrode. Different nano and micro-sized magnetic particles were evaluated in this approach. The 'IMS/m-GEC electrochemical immunosensing' system shows a limit of detection of 5×10^4 and 1×10^4 CFU mL^{-1} in BHI culturing media when micro and nanoparticles are used respectively. These LOD were achieved in a total assay time of 1 h without any previous culturing preenrichment step. Moreover, this system was able to clearly distinguish between food pathogenic bacteria such as *S. enterica* and *E. coli*. The features of this approach were discussed and compared with conventional culture methods.

1. Introduction

The increasing incidence of food-borne illnesses is becoming an important public health concern for consumers worldwide [1]. Conventional methods for food pathogen detection depend upon microbiological culturing techniques combined with standard biochemical identifications. These methods are time-consuming, laborious and might introduce sampling and enumeration errors due to the extremely low concentration of pathogenic bacteria in food samples able to produce disease [2].

In this context, electrochemical immunosensors presents a good advantage due to their high sensitivity, rapidity, low cost and possibility of being a hand-held platform for field applications [3].

Magnetic particles with sizes ranging from a few nanometres up to micrometer dimensions have been widely used for detection, concentration, separation, purification and identification of specific targeting ligands [4]. Their easily surface modification with functional organic molecular groups, such as amine, hydroxyl, carboxyl, aldehyde or thiol allows the immobilization of biomolecules (enzymes, antibodies, oligonucleotides or other biologically active compounds). Moreover, the possibility of being manipulated by an external magnetic field gradient provides a selective capture of target cells, offering an attractive technology for the design of biosensors systems, as well as microfluidic devices [5-7]. Recently in our group, a rapid strategy for the detection of *S. enterica* in milk combining immunomagnetic separation (IMS) and electrochemical magneto-immunosensing was developed,

[3] To whom any correspondence should be addressed.

using commercial anti-*Salmonella* particles [3]. In the present work, an electrochemical immunosensor for *Salmonella* detection with magnetic nano and micro-sized particles is reported. The strategy is based on the detection of the whole bacteria by a double immunological recognition. The effect of the size particle was evaluated by comparing nano and micro-sized magnetic particles modified with antibodies for the immunomagnetic separation (IMS) of the bacteria. Then, the captured bacteria were detected using a polyclonal anti-*Salmonella*-HRP antibody. The modified magnetic particles were easily captured by a magneto-electrode based on graphite-epoxy composite (m-GEC), which was also used as the transducer for the electrochemical detection. The features of this approach were discussed and compared with conventional culture methods.

2. Materials and methods

2.1. Chemicals and Biochemicals

Dynabeads M-280 Tosylactivated (tosyl-MMP, Product N° 142.04) and Anti-*Salmonella* magnetic beads (Product N° 710.02) were purchased from Dynal Biotech ASA (Oslo, Norway). Carboxyl-Adembeads (carboxyl-MNP, Product N° 0213) were supplied by Ademtech (Pessac, France). Mouse monoclonal Anti-*Salmonella* (Product N° ab 8274), rabbit polyclonal Anti-*Salmonella* (HRP) (Product N° ab 20771) and goat polyclonal secondary antibody to mouse IgG (HRP) (Product N° ab 20043) were purchased from Abcam (Cambridge, UK). The graphite-epoxy composite was prepared with graphite thin powder (1.04206.2500, Merck, Darmstadt, Germany) and Epo-Tek H77 (epoxy resin and hardener both from Epoxy Technology, USA). Hydrogen Peroxide and TMB (3,3',5,5'-tetrametylbenzidine) solutions used for optical measurements (TMB Substrate Kit, Reference no. 4834021) were purchased from Pierce (Rockford, USA).

Activation and storage buffers used for carboxyl-MNP were supplied from Ademtech (Pessac, France). All other buffer solutions were prepared with milli-Q water and all other reagents were in analytical reagent grade (supplied from Sigma and Merck). The composition of these solutions was: MES buffer (0.1 mol L^{-1} [2-(N-(morpholino)ethanesulfonic acid)], 0.9 % (w/v) sodium chloride, pH 4.7); borate buffer (0.1 mol L^{-1} H_3BO_3, pH 8.5); ammonium sulphate buffer (3 mol L^{-1} [$(NH_4)_2SO_4$] in borate buffer); phosphate blocking buffer (10 mmol L^{-1} sodium phosphate, 0.5% w/v BSA, pH 7.4); Ademtech blocking buffer (activation buffer, 0.5% w/v BSA); phosphate storage buffer (10 mmol L^{-1} sodium phosphate, 0.1% w/v BSA, 0.02 % (w/v) sodium azide, pH 7.4); PBS (10 mmol L^{-1}, pH 7.4); PBS 3% BSA (0.01 mol L^{-1}, pH 7.4, 3 % w/v BSA); sulphuric acid solution (2 mol L^{-1} H_2SO_4); PBST (10 mmol L^{-1} sodium phosphate, 0.8 % w/v NaCl, pH 7.4, 0.05 % v/v of Tween 20), PBST 2 % BSA (10 mmol L^{-1} sodium phosphate, 0.8 % w/v NaCl, pH 7.4, 0.05 % v/v of Tween 20, 2 % w/v BSA) and PBSE (0.1 mol L^{-1} sodium phosphate, 0.1 mol L^{-1} KCl, pH 7.0).

The bacterial strains used in this work were *S. enterica* serovar Typhimurium LT2 and *E. coli* DH5a. Both strains were grown at 37°C in Luria-Bertani (LB) agar plates. Bacterial suspensions were performed in Brain Heart Infusion (BHI) from freshly grown plates of the appropriated bacteria and its turbidity was adjusted at OD_{600} of 0.2. The exact concentration of each bacterial suspension was quantified by plating serial dilutions onto LB plates.

2.2. Instrumentation

All the incubations and washing steps were performed on microtiter plates under shaking conditions using a Minishaker MS1 (IKA, Germany). Temperature-controlled incubations with Eppendorf tubes were performed in an Eppendorf Thermomixer compact. Covalent immobilization of Anti-*Salmonella* antibody on the magnetic particles was performed using a rotor for test tubes (Micro Bio Tec-TTR 79). Magnetic separation during the washing steps was performed using a magnetic separator Dynal MPC-S (Product N° 120.20D, Dynal Biotech ASA, Norway). Polystyrene MaxiSorp microplates were purchased from Nunc (Catalogue no. 442404, Roskilde, DK). Optical measurements were performed on a TECAN Sunrise microplate reader with a Magellan v4.0 software. Amperometric measurements were performed with a LC-4C amperometric controller (BAS Bioanalytical Systems Inc, USA). A three electrode setup was used comprising a platinum auxiliary electrode (Crison 52-67 1), a double junction Ag/AgCl reference electrode (Orion 900200) with 0.1 mol L^{-1} KCl as the external reference solution and a working electrode (the magneto electrode, m-GEC). The detailed preparation of the m-GEC electrodes has been extensively described by Pividori *et al.* [8, 9].

2.3. Covalent immobilization of anti-Salmonella antibody on magnetic particles and coupling efficiency

Anti-*Salmonella* antibody was covalently coupled on both magnetic particles, tosyl-modified magnetic microparticles (tosyl-MMP, 2.8 μm, 30 mg mL^{-1}, 2 x 10^9 MP mL^{-1}) and activated carboxyl-modified magnetic nanoparticles (activated carboxyl-MNP, 300 nm, 30 mg mL^{-1}, 1 x 10^{12} MP mL^{-1}). Before antibody immobilization, carboxyl-MNP were activated as follows, 140 μL of carboxyl-MNP were resuspended and washed twice in 1 mL of activation buffer then, 4 mg mL^{-1} of 1-ethyl-3-(3-dimethylaminopropyl) carbodiimide (ECD) and 8 mg mL^{-1} of *N*-hydroxy-sulfosuccinimide (sulfo-NHS) were added at a final volum of 320 μL in MES buffer and incubated under shaking at room temperature during 10 h.

For the covalent immobilization, a volume of 140 μL of both tosyl-MMP and activated carboxyl-MNP were washed twice with 1 mL of borate buffer and activation buffer respectively, avoiding foaming. Afterwards, tosyl-MMP were resuspended in 560 μL of borate buffer and 400 μL of ammonium sulphate buffer, while activated carboxyl-MNP were resuspended in 960 μL of borate buffer. Then, 40 μL of the Anti-*Salmonella* antibody was added, reaching a concentration of 83.3 μg mL^{-1} in the immobilization solution. The MPs were incubated for a total reaction time of 13 h, 8 h at 37°C and 4 h at room temperature, with slow tilt rotation, using a rotor for the test tubes. After incubation, the supernatant was collected to perform the quantification of the remaining protein by ELISA immunoassay. Then, 1 mL of phosphate and Ademtech blocking buffers were added to tosyl-MMP and activated carboxyl-MNP respectively and incubated under shaking for 2 h at 37 °C, in order to inactivate the remaining tosyl and carboxyl molecular groups. Finally, the modified antibody-MPs were washed and resuspended in phosphate storage buffer (tosyl-MMP) or storage buffer (activated carboxyl-MNP) to reach a concentration of 5.0 mg mL^{-1} and stored at 4°C. Before each assay, MPs were washed twice and resuspended in PBST buffer.

In order to evaluate the antibody coupling efficiency for each magnetic particle, the concentration of antibody in the supernatant before and after the immobilization step was determined by ELISA immunoassay. In this context, serial dilutions of the supernatant collected during the covalent immobilization, were adsorbed on a 96-well microplate overnight at 4 °C. Two calibration curves were prepared with serial dilutions of anti-*Salmonella* antibody ranging from 0 to 0.15 μg mL^{-1} in 0.31 mol L^{-1} ammonium sulphate dissolved in 10 mmol L^{-1} PBS for tosyl-MMP and 24 mmol L^{-1} borate buffer for activated carboxyl-MNP. After two washing steps with 200 μL of phospathe buffer at room temperature and shaking for 5 min, plate blocking was performed under shaking with 200 μL of PBS 3% BSA, at room temperature for 2 h. Then, 3 washing steps with 200 μL of phosphate buffer were performed with shaking at room temperature for 5 min. The immobilized antibody was revealed with 100 μL of secondary antibody labeled with the enzyme HRP, diluted 1/3000 in PBST 2% BSA for 1 h, under shaking at room temperature. Afterwards, according to the kit manufacturer, a mixture of TMB/H_2O_2 in the proportion of 1:1 was reacted with the secondary peroxidase antibody, at room temperature with shaking for 30 min. The enzymatic reaction was stopped by adding 100 μL of H_2SO_4 (2 mol L^{-1}) Finally, the absorbance was measured at 450 nm after a short shaking and the standard curve was obtained plotting Abs vs. concentration (μg mL^{-1}) (Figure 2), for the quantification of anti-*Salmonella* attached on the magnetic particles.

2.4. Immunomagnetic separation. Evaluation by conventional culture methods

The schematic procedure for the immunomagnetic separation (IMS) of *S. enterica* is outlined in Figure 1, B. A volume of 10 μL (0.05 mg) of both anti-*Salmonella* modified magnetic particles (nano and micro-sized) and Dynabeads® anti-*Salmonella* was added to 500 μL of several serial dilutions of a *S. enterica* suspension ranged from a concentration of 4.42 x 10^5 to 4.42 x 10^1 CFU mL^{-1} in BHI broth.

An incubation step was performed for 30 min with slight agitation. After that, the bacteria attached to the particles were separated with a magnet and the supernatant was collected for plating. Then, MPs were washed with PBST (x5) shaking for 1 min at room temperature. Finally, the collected modified MPs were resuspended in 110 μL of PBST. Both, MPs with the attached bacteria and supernant were plated on LB agar and grown for 18-24 h at 37 °C.

2.5. Electrochemical magneto-immunosensing

The IMS/m-GEC electrochemical immunosensing procedure is schematically outlined in Figure 1, A, and it is based on the following steps: (i) immunomagnetic separation (IMS); (ii) immunological reaction with the anti-*Salmonella*-HRP antibody; (iii) electrochemical detection. The IMS step was performed as described in Section 2.4. For each concentration of bacteria ranged from 0 to 3.70 x 10^5 CFU mL^{-1} for carboxyl-MNP and from 0 to 5.48 x 10^5 CFU mL^{-1} for tosyl-MMP in BHI broth. After discarding the supernatant, the collected MPs were further reacted with 140 µL of Anti-*Salmonella* HRP antibody (diluted 1/1000 in PBST 2% BSA) for 30 min at room temperature. Finally, the magnetic particles were washed twice and resuspended in 140 uL of PBST. After the final washing step, the modified magnetic beads were captured by dipping the magneto electrode (m-GEC) inside the reaction tube. The modified m-GEC electrode was immersed into the electrochemical cell containing 20 mL of PBSE buffer with 1.81 mmol L^{-1} hydroquinone, and under continuous magnetic stirring, a potential of -0.100 V vs Ag/AgCl was applied. When a stable baseline was reached, 500 μL of H_2O_2 was added into the electrochemical cell to a final concentration of 4.90 mmol L^{-1} (which corresponds to the H_2O_2 concentration capable to saturate the whole enzyme amount employed in the labeling procedure), and the current was measured until the steady state current was reached. This steady-state current was used for Figures 3 and 4.

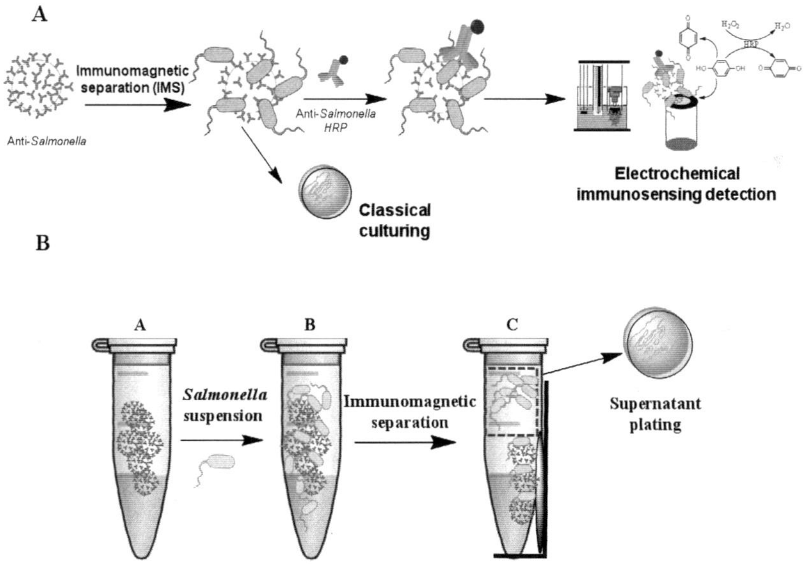

Figure 1. Schematic representation of electrochemical immunosensing procedure for *S. enterica* (up panel), as well as for the evaluation for the immunomagnetic separation (IMS) by conventional culture methods (down panel).

2.6. Specificity study

In order to verify the specificity of this approach, the above procedure was also performed using carboxyl-MNP with 5.90 x 10^4 CFU mL^{-1} of *E. coli*, 1.85 x 10^5 CFU mL^{-1} of *S. enterica* and, finally, a sample containing both bacterial species (5.90 x 10^4 and 1.85 x 10^5 CFU mL^{-1}, respectively for *E. coli* and *S. enterica*) artificially inoculated in BHI. The specificity was studied for tosyl-MMP as well, testing 4.90 x 10^4 CFU mL^{-1} of *E. coli*, 2.74 x 10^5 CFU mL^{-1} of *S. enterica* and, finally, a sample containing both bacterial species (4.90 x 10^4 and 2.74 x 10^5 CFU mL^{-1}, respectively for *E. coli* and *S. enterica*) artificially inoculated in BHI. Two negative controls were also performed for each type of magnetic particle.

3. Results and Discussion

Covalent immobilization efficiency of anti-*Salmonella* antibody on magnetic particles was performed by quantifying the amount of antibody present in the supernatant before and after the immobilization step through an ELISA immunoassay. Figure 2 shows the calibration curves for both types of magnetic particles, presenting a coupling efficiency of 90 and 99% of antibody immobilized on the magnetic particles, for nanoparticles (carboxyl-MNP) and microparticles (tosyl-MMP), respectively.

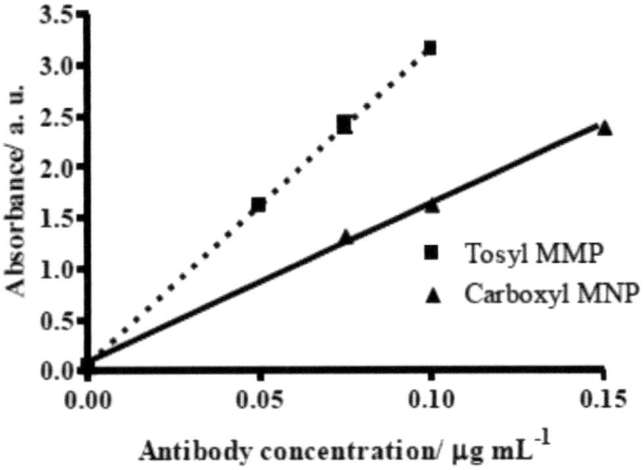

Figure 2. Calibration curve for tosyl-MNP in 0.31 mol L^{-1} ammonium sulphate dissolved in 0.10 mmol L^{-1} phosphate buffer and carboxyl-MNP in 24 mmol L^{-1} borate buffer for ELISA immunoassay. The error bars show the standard deviation for n = 3.

The immunomagnetic separation step (IMS) was evaluated by conventional culture methods for the modified magnetic particles, as well as for the commercial anti-*Salmonella* particles. Table 1 shows the bacteria capture efficiency obtained by culturing MPs after the IMS. Similar capture efficiencies were obtained for all magnetic particles, showing promising capability for the further electrochemical immunosensing.

Table 1. Immunomagnetic separation of *Salmonella* with Dynabeads® anti-*Salmonella* and nano and micro-sized magnetic particles.

S. enterica concentration/ CFU mL^{-1}	Immunomagnetic separation (IMS)/ %		
	Dynabeads® anti-*Salmonella*	Carboxyl-MNP	Tosyl-MMP
10^5	**92**	**85**	**90**
10^4	**77**	**73**	**85**
10^3	**90**	**95**	**83**
10^2	**82**	**95**	**95**
10^1	**100**	**100**	**100**

The IMS/m-GEC electrochemical immunosensing procedure was developed for both magnetic particles (tosyl-MMP and carboxyl-MNP). The amperometric signal corresponding to the LOD was estimated by processing the negative control samples of 0 CFU mL^{-1} in BHI and performing three different single inter-day assays, using ten different magneto electrode devices, obtaining a mean value of 0.32 μA with a standard deviation of 0.07 μA for the tosyl-MMP and a mean value of 0.51 μA with a standard deviation of 0.08 μA for the carboxyl-MNP. The amperometric signal corresponding to the LOD value was then extracted with a one-tailed t-test at a 99 % confidence level, giving a value of 0.53 and 0.73 μA respectively for tosyl-MMP and carboxyl-MNP (shown in Figure 3 as the dotted and solid horizontal lines, respectively). The results obtained show that IMS/m-GEC electrochemical immunosensing approach was able to detect 5 x 10^4 CFU mL^{-1} (tosyl-MMP) and 1 x 10^4 CFU mL^{-1} (carboxyl-MNP) of *S. enterica* in BHI broth, in 1 h.

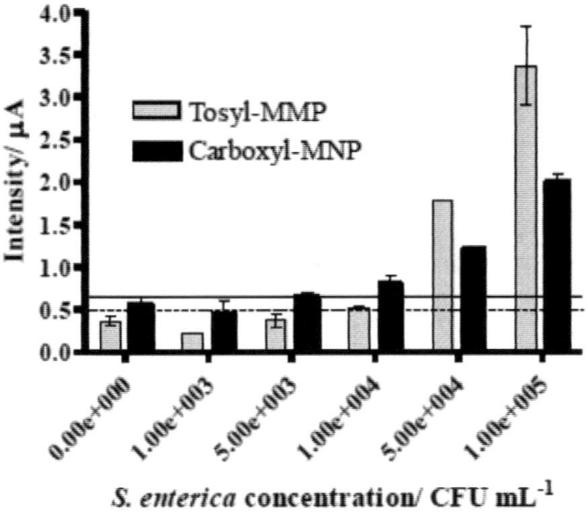

Figure 3. 'IMS/m-GEC electrochemical immunosensing' approach for the detection from 10^3 to 10^5 CFU mL^{-1} of *S. enterica* cells artificially inoculated in BHI broth. The error bars show the standard deviation for n = 2, except for the negative control (n = 9 for Carboxl-MNP and n = 7 for Tosyl-MMP in BHI broth. The LOD values for tosyl-MMP (dotted line) and carboxy-MNP (solid line) are also displayed.

Figure 4 shows the results obtained for the specificity study. As expected, the electrochemical signal obtained for *E. coli* was similar to the value obtained when processing the negative control, while the mix of both pathogens (*E. coli* and *S. enterica*) gave a similar signal than the sample spiked just with *S. enteric* for both particles nano and micro-sized. These results confirm the specificity of the 'IMS/m-GEC electrochemical immunosensing' approach.

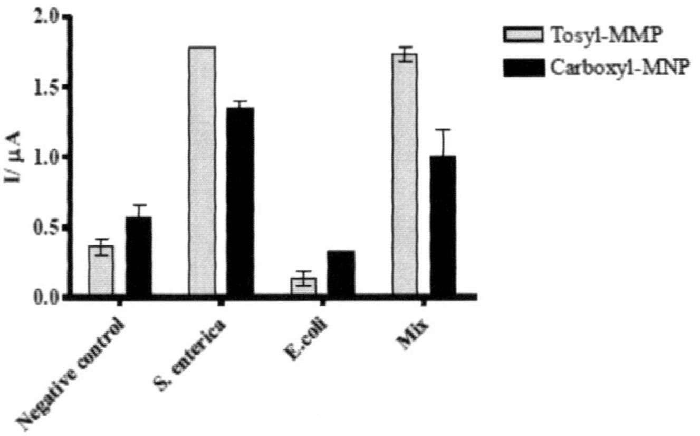

Figure 4. Specificity study for the electrochemical immuno-sensing approach. The bars show the electrochemical signal for BHI broth with: 0 CFUmL^{-1} (negative control); 5.9×10^4 CFU mL^{-1} of *E. coli*; 1,85×10^5 CFU mL^{-1} of *S. enterica*; and a mix solution containing 5.9×10^4 CFU mL^{-1} of *E. coli* and 1,85×10^5 CFU mL^{-1} of *S. enterica*. The error bars show the standard deviation for n=2.

4. Conclusions

A specific and sensitive electrochemical magneto immunosensor for *S. enterica* detection was developed. Evaluation of different magnetic particles (nano and micro-sized) functionalized with different molecular groups showed to be very efficient on the IMS of *S. enterica*, which capture was achieved as low as 10 CFU mL^{-1} by conventional culturing methods. Furthermore, the use of nanoparticles improved the IMS/m-GEC electrochemical immunosensing approach being able to detect as low as 1 x 10^4 CFU mL^{-1} in BHI broth, in 1 h without any culturing enrichment. Magnetic nanoparticles offer a good advantage for food pathogenic bacteria detection, presenting a rapid alternative to the laborious and time consuming culturing detection methodologies. The high sensitivity of the approach conferred by the m-GEC electrochemical immunosensing coupled with magnetic separation using magnetic nanoparticles, results in an extremely specific, rapid, robust and sensitive procedure, all of them promising features for being implemented as microfluidic system mainly for food industry applications. Future work will focus on further validation of this assay in artificially inoculated as well as in naturally contaminated meats, poultry, dairy products, and environmental samples.

Acknowledgments

Financial support from BioMaX: Novel diagnostic bio-assays based on magnetic particles Marie Curie Initial Training Networks (ITN) Call: FP7-PEOPLE-2010-ITN, Ministry of Science and Innovation (MEC), Madrid (Project BIO2010-17566) and from Generalitat de Catalunya (Projects SGR 323 and SGR2009-1106) are acknowledged. Ademtech (Pessac, France) is also acknowledged for providing Carboxyl-Adembeads magnetic nanoparticles.

5. References

[1] Suo B, He Y, Paoli G, Gehring A, Tu S, Shi X 2010 *Mol. Cell. Probes* **24** 77–86
[2] Liébana S, Lermo A , Campoy S, Barbé J, Alegret S, Pividori M I 2009 *Anal. Chem.* **81** 5812–5820
[3] Liébana S, Lermo A , Campoy S, Barbé J, Alegret S, Pividori M I 2009 *Biosens. Bioelectron.* **25** 510-513
[4] Roque A C A, Bispo S, Pinheiro A R N, Antunes J M A, Gonçalves D, Ferreira H A 2009 *J. Mol. Recognit.* **22** 77–82
[5] Hsing I-Ming, Xu Y, Zhao W 2007 *Electroanalysis* **19** 755 – 768
[6] Yang H, Li H, Jiang X 2008 *Microfluid. Nanofluidics* **5** 571–583
[7] Den Dulk R C, Schmidt K A, Sabatté G, Liébana S, Prins M W J, 2013 *Lab Chip* **13** 106-118
[8] Pividori M I, Alegret S, 2005 *Anal Lett* **38** 2541-2565
[9] Pividori M I, Lermo A, Campoy S, Barbe J, Alegret S 2007 *Electrochemical Sensor Analysis* Elsevier Amsterdam 221-226

Journal of Physics: Conference Series **421** (2013) 012021 doi:10.1088/1742-6596/421/1/012021

Development of an electrochemical biosensor for the detection of an ADP-ribosylating toxin, exo A from *Pseudomonas aeruginosa*

Y Enríquez, Y Negrón, M Navarreto and A R Guadalupe[1]
University of Puerto Rico, Río Piedras Campus, Department of Chemistry, PO Box 23346, San Juan, PR 00931-3346

Email: anlupe55@gmail.com

Abstract. A free radical copolymerization of Styrene (Sty) and acrylic acid N-hydroxysuccinimide ester (NAS) has been done in a range of 10:90 to 90:10 (Sty:NAS) molar ratios. The FT-IR spectra for all seven copolymers showed the absorption peaks for the carbonyl signals of the ester and the amide in NAS (1773 cm^{-1} and 1727 cm^{-1} respectively), the styrene aromatic signal (1494 cm^{-1}) and the disappearance of the absorption peak for the vinyl group in both monomers (1629 cm^{-1}). HPLC-UV results showed an increment in the average molecular weight with an increase in the molar ratio of the styrene monomer, from 1528.51 g/mol for 10:90 Sty:NAS to 7141.67 g/mol for 90:10 Sty:NAS. These copolymers will be used to generate films on carbon surfaces to anchor a β -NAD^+ electroactive analog. Also, a Ferrocene-labeled NAAD (Fc-NAAD) was prepared by attaching Ferrocene Succinimide (Fc-NHS) to the primary amine in the adenine moiety of the cofactor. Osteryoung Squatre Wave Voltammetry (OSWV) of the new Fc-NAAD showed an anodic peak in 320 mV and the cyclic voltammetry (CV) showed chemical reversibility and electrochemical quasi-reversibility.

1. Introduction

Pseudomonas aeruginosa is an opportunistic pathogen and is considered one of the most common bacterium found in nosocomial infections [1,2]. The capacity of *Pseudomonas aeruginosa* (PA) to infect a tissue depends upon the production of virulence factors like enzymes and toxins that damage the host cell [3,4]. Exotoxin A has been reported to be the most toxic protein secreted by *Pseudomonas aeruginosa* [4,5]. This exotoxin catalyzes the transfer of the ADP-ribose of β-NAD^+ to eukaryotic elongation factor 2 (eEF-2) [6,7,8]. This ADP-ribosylation of eEF-2 affects the protein synthesis process at the polypeptide chain elongation resulting in cell death [6,7,8]. The selective identification of deathly and viable pathogens would be a huge advantage in biotechnology and biomedical research. Their early detection will be the key to a timely treatment for patients.

Our intention is to detect Exo A thru the ADP-ribosylation of eEF-2 in the presence of ferrocene-labeled NAAD. To achieve this, we synthesized and characterized seven Sty:NAS copolymers at different molar ratios and a Fc-NAAD cofactor as the first step in the development of an electrochemical biosensor for the detection of Exo A. The copolymers will be used for the modification of a glassy carbon electrode surface and the attachment of the Fc-NAAD through a

[1] To whom any correspondence should be addressed.

diamine linker. We expect to measure a decrease in the initial ferrocene redox signal after the occurrence of the eEF2 ribosylation reaction.

2. Methodology

2.1. Synthesis of a Styrene: N-acryloxysuccinimide copolymer (Scheme 1)

A free radical copolymerization of seven copolymers of Sty:NAS with different molar ratios (10:90, 25:75, 40:60, 50:50, 60:40, 75:25, 90:10) was done using 2-2'-Azoisobutyronitrile (AIBN) as the initiator. The polymerization was done under N_2 (g) reflux at 70 ^{0}C – 75 ^{0}C for 24 hours. The final product was purified in methanol and characterized by FT-IR and HPLC-UV.

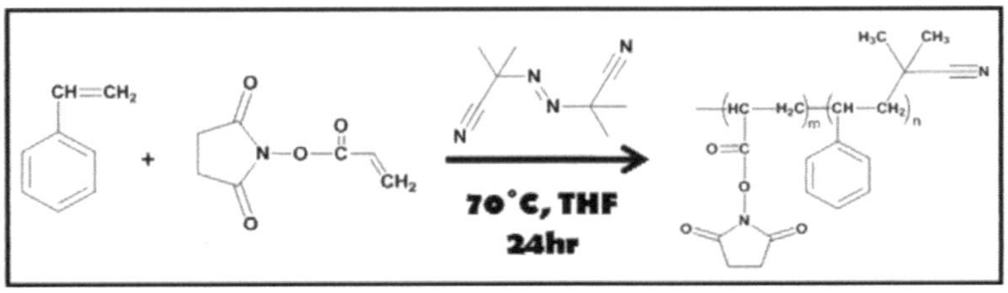

Scheme 1. Synthesis of Styrene and NAS copolymer

2.2. Synthesis of a Ferrocene labeled NAAD (Scheme 2)

A Ferrocene-labeled NAAD cofactor (Fc-NAAD) was prepared by attaching Ferrocene carboxylic N-hydroxysuccinimide ester (Fc-NHS) to the primary amine in the adenine moiety of nicotinic acid adenine dinucleotide (NAAD). The NAAD and Fc-NHS solutions were mixed in a 1:1.2 molar ratio respectively and the synthesis was done at 25 °C. The final product was purified using a column of Sephadex G-15.

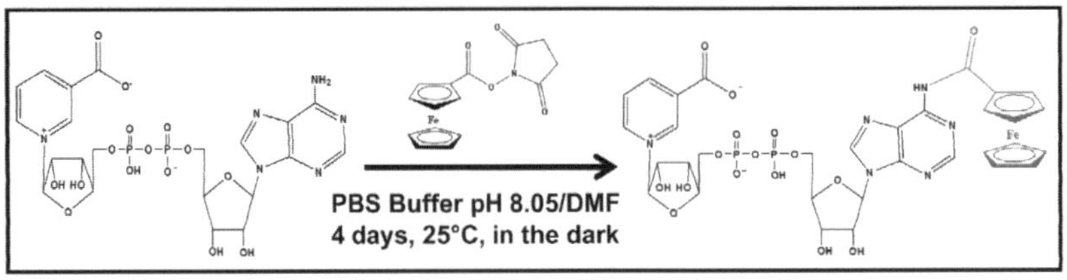

Scheme 2. Synthesis of Fc-NAAD using Fc-NHS.

3. Results and Discussion

3.1. Sty:NAS copolymer synthesis and characterization

Figure 1 shows the FT-IR for seven Sty:NAS copolymers and Table 1 summarizes the FT-IR results for all copolymers. FT-IR analysis was done with Buker Tensor 27 equipped with a diamond point and OPUS data collection program. For the Styrene FT-IR we used a NaCl sealed liquid cell.We assigned the 1727-1738 cm^{-1} signal to the presence of the C=C in the styrene aromatic group and the 1773-1781 cm^{-1} signal to the carbonyl in the ester functionality of NAS. We also observed the symmetrical and antisymmetrical vibration modes of the carbonyl in the succinimide [9] around 1773-1810 cm^{-1}. Furthermore, in the copolymers spectra we observed the disappearance of the vinyl signal around 1629 cm^{-1} suggesting the formation of a covalent bond between the monomers.

Table 1. Experimental values for the FT-IR frequencies of the Sty:NAS copolymers at different molar ratios.

	Styrene	Acrylic Acid N-hydroxysuccinimide ester		
Copolymers Sty:NAS	C=C Aromatic stretch Experimental Value Styrene: 1494.77 cm^{-1}	Ester C=O stretch Experimental Value NAS: 1728.14 cm^{-1}	NHS C=O stretch Experimental Value NAS: 1774.43, 1797.58 cm^{-1}	
10:90	1494.75	1731.57	1778.88	1809.68
25:75	1494.28	1733.79	1779.28	1809.59
40:60	1494.32	1734.64	1779.28	1809.26
50:50	1494.13	1734.66	1779.17	1809.10
60:40	1493.80	1735.32	1779.07	1808.87
75:25	1493.79	1736.45	1779.96	1809.17
90:10	1492.92	1738.28	1780.60	1809.68

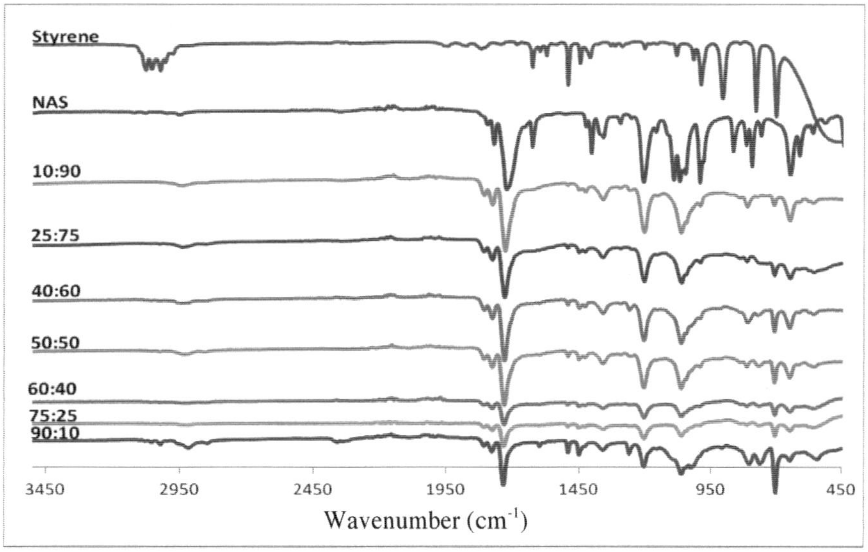

Figure 1. Experimental results for the FT-IR of seven copolymers of Sty:NAS with different molar ratios (10:90, 25:75, 40:60, 50:50, 60:40, 75:25, 90:10) compared with Styrene and acrylic acid N-hydroxysuccinimide ester.

The molecular weight (Mn and Mw) of each copolymer was determined by Size Exclusion Chromatography (SEC) [Hewlett Packard 1100 Series]. The HPLC system was equipped with the UV spectrophotometer (Wyatt Technology Corporation) and two Phenogel colums (Phenomenex) with effective molecular weight ranges of 1,000 – 7,500 g/mol and 5,000 – 500,000 g/mol. Using a third order polynomial curve (Figure 2) we determined that the Mn of the copolymers increases from 1528.51 g/mol to 7141.67 g/mol while the molar ratio of styrene/NAS increases from 10:90 to 90:10 (Sty:NAS) [Table 2]. The 50:50 (Sty:NAS) showed the higher yield of product with a 30.11% and the 90:10 (Sty:NAS) the lowest with a 15.04%.

Table 2. HPLC-UV Experimental Data of Mn, Mw and Mw/Mn for (Sty:NAS) Copolymers.

Copolymer (Sty:NAS)	Experimental Mn (g/mol)	Experimental Mw (g/mol)	Polydispersity Mw/Mn
10:90	1528.51	1644.51	1.08
25:75	3676.17	4065.91	1.11
40:60	4770.91	5248.25	1.10
50:50	6215.55	6749.13	1.09
60:40	5605.39	6122.50	1.09
75:25	5926.16	6453.11	1.09
90:10	7141.67	7685.23	1.08

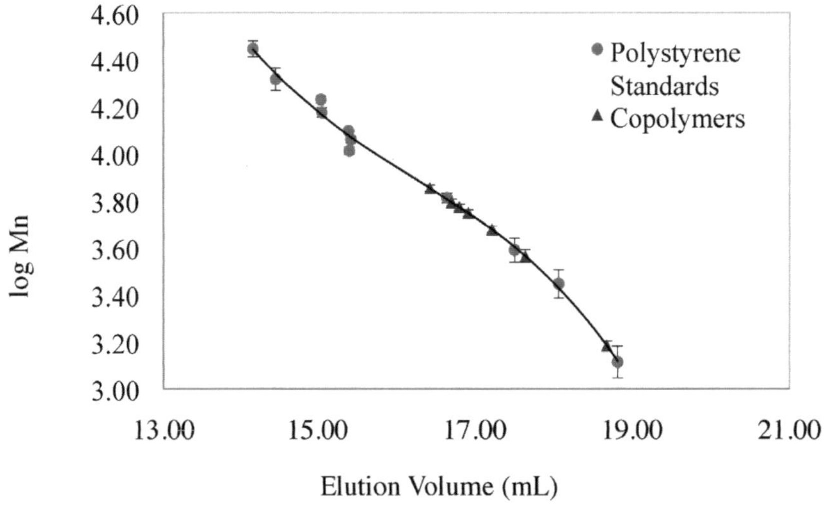

Figure 2. Polynomial calibration curve for Polystyrene Standards and Sty:NAS copolymers in THF.

3.2. Fc-NAAD synthesis and Characterization

^{1}HNMR confirmed the formation of the Fc-NAAD with chemical shifts for the Ferrocene moiety δ(s)4.222, δ(s)4.373, δ(s)4.486, the pyridine ring in NAAD δ(s)8.094, δ(s)8.383 and the adenine ring in NAAD δ(d)8.693;8.709, δ(d)8.940;8.952, δ(s)9.089. One dimensional NMR analysis was done in Deuterium Oxide (D_2O) with the 500 MHz Bruker Tensor. For the electrochemical characterization we used the Bioanalytical System Epsilon electrochemical analyzer and a three electrode cell system with glassy carbon as the working electrode, Ag/AgCl (NaCl, 3M) as the reference electrode and Nichrome, as the counter electrode. Cyclic Voltammetry (CV) was done using a 1.116 x 10^{-3} M solution of Fc-NAAD in 1:3 acetonitrile:0.0201 M PBS (pH = 7.10, μ= 0.10M). The potential window was 0-700 mV going in the positive direction at a scan rate of 50, 100, 200, 500 and 1,000 mV/s (Figure 3).

The current increased with higher scan rate (0.8 ± 0.4 at 50 mV/s to 4.7 ± 0.9 at 1,000 mV/s) and a formal redox potential of (327 ± 2) mV at 200 mV/s was observed (Figure 3a). Our new product shows chemical reversibility (Ipa/Ipc = 0.95 ± 0.24 A at 200 mV/s) and electrochemical quasi-reversibility. A Log Ipa vs Log v curve showed a slope of 0.5812 and a linearity of R^2 = 0.9987, a typical behavior for a diffusion controlled electrochemical process. OSWV analysis showed an anodic peak in 328 mV for Fc-NAAD. Compared with the anodic peak of Fc-NHS in 656 mV the potential decreases for Fc-NAAD.

a) b)

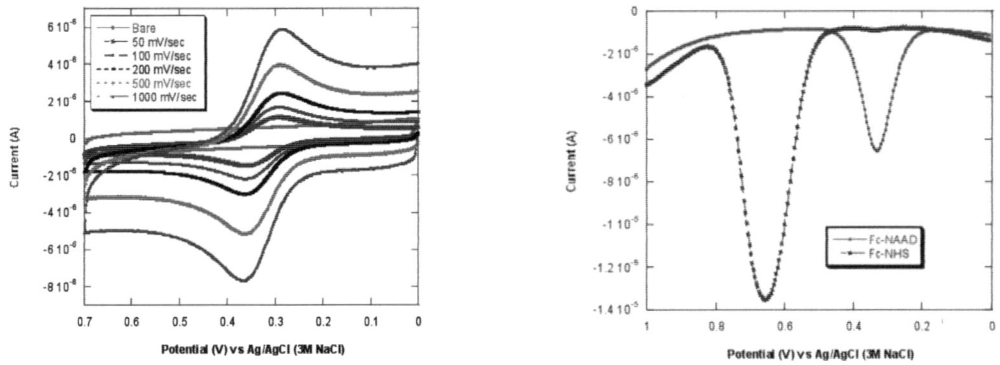

Figure 3. (a) Cyclic voltammetry at different scan rates for 1.116 x 10^{-3} M Fc-NAAD in 1:3 CH_3CN: 0.0201 M PBS (pH = 7.10, μ= 0.10M). (b) Osteryoung square wave voltammetry of Fc-NAAD and Fc-NHS in 1:3 CH_3CN: 0.0201 M PBS (pH = 7.10, μ= 0.10M).

4. Conclusion

The synthesis and characterization of Sty:NAS copolymer was done successfully for the different molar ratios (10:90 – 90:10). FT-IR spectra for all seven copolymers showed the absorption peaks for the carbonyl signals of the ester and the amide in NAS, the styrene aromatic signal and the disappearance of the absorption peak for the vinyl group in both monomers. HPLC-UV results showed an increase in the average molecular weight with increase in the molar ratio of the styrene/NAS and the 50:50 Sty:NAS copolymer had the highest yield, 30.11%. Fc-NAAD was prepared by attaching Fc-NHS to the primary amine in the adenine moiety of the cofactor. CV analysis showed a current increase with higher scan rate and a formal redox potential of (327 ± 2) mV at 200 mV/s. Fc-NAAD showed chemical reversibility and electrochemical quasi-reversibility.

Our work is now focused on the full characterization of the Fc-NAAD cofactor. The carboxyl group in the nicotinic acid moiety of the Fc-NAAD will be used for its immobilization on the electrode surface through an amine linker and our final goal is the electrochemical monitoring of the ADP-ribosylation process mediated by Exo A in vitro.

Acknowledgments

We acknowledge the kind support of RISE Program Award ID: 2R25GM061151-11and PENN-UPRH Partnership for Research and Education (NSF-DMR- 0934195).

5. References

[1] Val Delden C and Iglewski B 1998 *Emerging Infectious Diseases* ***4*** 551
[2] Cornelis P. 2008 *Caister Academic Press*, UK 129
[3] Salyers A and Whitt D 1994 *Bacterial Pathogenesis: A molecular approach* 260-9
[4] Iglewski B, Sadoff J, Bjorn M and Maxwell E 1978 *Proceedings of the National Academy of Science* **75** 3211
[5] Wick M, Hamood A and Iglewski B 1990 *Molecular Microbiology* **4** 527
[6] Leppla S Large 1976 *Infection and Immunity* **14** 1077
[7] Jorgensen R, Merrill A, Yates S, Marquez V and Schwan A 2005 *Nature* **436** 979
[8] Pallen M, Lam A, Loman N and McBride A 2001 *Trends in Microbiology* **9** 302
[9] Hargreaves M, Pritchard J and Dave H 1970 *Chemical Reviews* **70** 439

AUTHOR INDEX

Aceves-Mijares, M 90
Aguilar-Muñiz, A U 96
Alegret, S 116
Almeida, F L 47
Aragón, R 1
Araki, K 26
Armas, L E G 26
Armentano, R L 101
Balay, G 101
Blanco, A 85
Brandão, D 116
Camargo, C J 5
Campoy, S 116
Cardona, R A 35
Cerize, N P N 68
Chávez, C 90
Colom-Ustáriz, J G 80
Cortés, P 116
Crand, S 31
Cymberknop, L 101
Da Silva, A N R 73
Da Silva, M L P 73
De Sagazan, O 31
Derenzo, S 68
Díaz-Serrano, M 42
Durán, A 85
Enríquez, Y 123
Esteve, J 5
Fachini, E R 11, 73
Fernandes, F J R 26
Fernández, D G 85
Filho, S G Dos S 47
Fontes, M B A 47, 109
Fuentes, O A-de 85
Garção, A S 61
García-Vargas, G G 96
Góngora-Rubio, M R 68
Guadalupe, A R 42, 123
Hernandez, L F 11
Huila, M F G 26
Jiménez-Jorquera, C 85
Leite, A R 11
Liébana, S 116
Lima, R R 11
Louro, P 61
Maioco, S 1
Melendez, A 5
Mohammed-Brahim, T 31
Morimoto, N I 31, 55
Navarreto, M 123
Negreira, C A 101
Negrón, Y 123
Oliveira, A M 68
Ortiz, J A 80
Park, J W 35
Pedraza, J 90
Peres, H E M 26
Pinto, N J 20
Pividori, M I 116
Pojar, M 26
Rajchenberg, N 1
Ramírez, J M 90
Ramos, I 5
Robles, J 5
Rodrigues, B S 31, 55
Román-López, S 90
Rosado, A 42
Sahoo, H 35
Santana, D 42
Santiago, N G 80
Santiago-Avilés, J 35
Santos, A D 26
Schianti, J N 68
Seabra, A C 26
Siarkowski, A L 31, 55
Silva, M L P 11
Silva, V 61
Suárez-Bagnasco, D 101
Toma, H E 26
Valdés-Perezgasga, F 96
Valle, M A 26
Vega, E Z 42
Vera, C 1
Vieira, M 61, 61

Institute of Physics
Dirac House, Temple Back
Bristol BS1 6BE UK

CURRAN ASSOCIATES INC
proceedings.com